AF812446

DIVISION FAST FACTS

Timed Math Facts Practice Tests

This book belongs to:

Welcome to *Division Fast Facts!* This workbook is designed to help you master division by covering all facts for dividing by 1-12 using 100 timed tests. By practicing regularly, you'll strengthen your division skills, increase your speed, and build the confidence you need to succeed in math.

How to Use This Workbook:

- Set a Timer: Each test is designed to be completed as quickly as possible. Use a timer to track your speed and challenge yourself to improve.
- Practice Regularly: Consistency is key! Work on one or two tests each day to steadily improve your skills.
- Review Mistakes: After each test, check the answer key in the back, review any errors, and focus on the facts that need extra practice.
- Track Your Progress: At the top of each page is space to record your time and score. In the back of the book, you will also find a progress tracking chart you can use to record your growth over time. Use it to see how far you've come and to set goals for improvement. Remember, progress takes time—keep practicing, and you'll see great results!

Good luck, and have fun mastering your division facts!

Copyright © 2024 EDventure Learning LLC

All rights reserved. This book or any portion thereof may not be reproduced or used in any manner whatsoever without the express written permission of the publisher.

Printed in the United States of America
ISBN:978-1-64824-046-1
EDventure Learning LLC
5601 State Route 31 #1296
Clay, NY 13039

www.edventurelearning.com

Email us at hello@edventurelearning.com
Follow us on social media @edventurelearning

CONTENTS

1

Divide by 1

Time: ____________

Score: ______/ 40

(1) $64 \div 1$	(2) $5 \div 1$	(3) $1 \div 1$	(4) $91 \div 1$	(5) $74 \div 1$
(6) $14 \div 1$	(7) $7 \div 1$	(8) $98 \div 1$	(9) $46 \div 1$	(10) $32 \div 1$
(11) $45 \div 1$	(12) $67 \div 1$	(13) $38 \div 1$	(14) $81 \div 1$	(15) $27 \div 1$
(16) $95 \div 1$	(17) $62 \div 1$	(18) $40 \div 1$	(19) $19 \div 1$	(20) $37 \div 1$
(21) $71 \div 1$	(22) $16 \div 1$	(23) $83 \div 1$	(24) $97 \div 1$	(25) $86 \div 1$
(26) $75 \div 1$	(27) $42 \div 1$	(28) $58 \div 1$	(29) $77 \div 1$	(30) $33 \div 1$
(31) $84 \div 1$	(32) $26 \div 1$	(33) $18 \div 1$	(34) $80 \div 1$	(35) $3 \div 1$
(36) $85 \div 1$	(37) $20 \div 1$	(38) $68 \div 1$	(39) $22 \div 1$	(40) $70 \div 1$

2

Divide by 1

Time: ___________

Score: _____/ 40

(1) 39 ÷ 1	(2) 88 ÷ 1	(3) 52 ÷ 1	(4) 57 ÷ 1	(5) 0 ÷ 1
(6) 92 ÷ 1	(7) 65 ÷ 1	(8) 61 ÷ 1	(9) 72 ÷ 1	(10) 29 ÷ 1
(11) 63 ÷ 1	(12) 24 ÷ 1	(13) 25 ÷ 1	(14) 93 ÷ 1	(15) 28 ÷ 1
(16) 21 ÷ 1	(17) 50 ÷ 1	(18) 89 ÷ 1	(19) 90 ÷ 1	(20) 87 ÷ 1
(21) 35 ÷ 1	(22) 9 ÷ 1	(23) 69 ÷ 1	(24) 82 ÷ 1	(25) 79 ÷ 1
(26) 76 ÷ 1	(27) 41 ÷ 1	(28) 56 ÷ 1	(29) 31 ÷ 1	(30) 96 ÷ 1
(31) 47 ÷ 1	(32) 30 ÷ 1	(33) 36 ÷ 1	(34) 59 ÷ 1	(35) 44 ÷ 1
(36) 94 ÷ 1	(37) 23 ÷ 1	(38) 4 ÷ 1	(39) 53 ÷ 1	(40) 73 ÷ 1

3

Divide by 1

Time: ___________

Score: _____/ 40

1. $78 \div 1$
2. $49 \div 1$
3. $10 \div 1$
4. $15 \div 1$
5. $99 \div 1$
6. $17 \div 1$
7. $13 \div 1$
8. $8 \div 1$
9. $2 \div 1$
10. $43 \div 1$
11. $55 \div 1$
12. $11 \div 1$
13. $100 \div 1$
14. $66 \div 1$
15. $51 \div 1$
16. $60 \div 1$
17. $12 \div 1$
18. $34 \div 1$
19. $48 \div 1$
20. $54 \div 1$
21. $6 \div 1$
22. $17 \div 1$
23. $38 \div 1$
24. $95 \div 1$
25. $66 \div 1$
26. $4 \div 1$
27. $42 \div 1$
28. $21 \div 1$
29. $80 \div 1$
30. $88 \div 1$
31. $34 \div 1$
32. $71 \div 1$
33. $93 \div 1$
34. $39 \div 1$
35. $30 \div 1$
36. $32 \div 1$
37. $36 \div 1$
38. $89 \div 1$
39. $48 \div 1$
40. $15 \div 1$

4

Divide by 1

Time: ___________

Score: _____/ 40

(1) 100 ÷ 1	(2) 9 ÷ 1	(3) 97 ÷ 1	(4) 50 ÷ 1	(5) 72 ÷ 1
(6) 44 ÷ 1	(7) 61 ÷ 1	(8) 77 ÷ 1	(9) 58 ÷ 1	(10) 98 ÷ 1
(11) 84 ÷ 1	(12) 35 ÷ 1	(13) 7 ÷ 1	(14) 26 ÷ 1	(15) 91 ÷ 1
(16) 25 ÷ 1	(17) 55 ÷ 1	(18) 86 ÷ 1	(19) 64 ÷ 1	(20) 0 ÷ 1
(21) 92 ÷ 1	(22) 78 ÷ 1	(23) 29 ÷ 1	(24) 63 ÷ 1	(25) 53 ÷ 1
(26) 52 ÷ 1	(27) 11 ÷ 1	(28) 74 ÷ 1	(29) 76 ÷ 1	(30) 14 ÷ 1
(31) 90 ÷ 1	(32) 83 ÷ 1	(33) 8 ÷ 1	(34) 12 ÷ 1	(35) 3 ÷ 1
(36) 31 ÷ 1	(37) 10 ÷ 1	(38) 23 ÷ 1	(39) 46 ÷ 1	(40) 47 ÷ 1

5

Divide by 2

Time: ________

Score: _____/ 40

(1) 12 ÷ 2	(2) 16 ÷ 2	(3) 6 ÷ 2	(4) 0 ÷ 2	(5) 10 ÷ 2
(6) 14 ÷ 2	(7) 22 ÷ 2	(8) 4 ÷ 2	(9) 18 ÷ 2	(10) 2 ÷ 2
(11) 20 ÷ 2	(12) 8 ÷ 2	(13) 24 ÷ 2	(14) 22 ÷ 2	(15) 4 ÷ 2
(16) 14 ÷ 2	(17) 10 ÷ 2	(18) 0 ÷ 2	(19) 18 ÷ 2	(20) 16 ÷ 2
(21) 6 ÷ 2	(22) 12 ÷ 2	(23) 8 ÷ 2	(24) 20 ÷ 2	(25) 24 ÷ 2
(26) 2 ÷ 2	(27) 16 ÷ 2	(28) 24 ÷ 2	(29) 0 ÷ 2	(30) 8 ÷ 2
(31) 20 ÷ 2	(32) 10 ÷ 2	(33) 12 ÷ 2	(34) 14 ÷ 2	(35) 4 ÷ 2
(36) 18 ÷ 2	(37) 2 ÷ 2	(38) 22 ÷ 2	(39) 6 ÷ 2	(40) 14 ÷ 2

6

Divide by 2

Time: ____________

Score: _____/ 40

(1) 0 ÷ 2	(2) 16 ÷ 2	(3) 22 ÷ 2	(4) 10 ÷ 2	(5) 20 ÷ 2
(6) 4 ÷ 2	(7) 8 ÷ 2	(8) 12 ÷ 2	(9) 24 ÷ 2	(10) 18 ÷ 2
(11) 6 ÷ 2	(12) 2 ÷ 2	(13) 12 ÷ 2	(14) 2 ÷ 2	(15) 14 ÷ 2
(16) 10 ÷ 2	(17) 16 ÷ 2	(18) 8 ÷ 2	(19) 18 ÷ 2	(20) 0 ÷ 2
(21) 24 ÷ 2	(22) 20 ÷ 2	(23) 4 ÷ 2	(24) 6 ÷ 2	(25) 22 ÷ 2
(26) 12 ÷ 2	(27) 24 ÷ 2	(28) 4 ÷ 2	(29) 20 ÷ 2	(30) 16 ÷ 2
(31) 22 ÷ 2	(32) 6 ÷ 2	(33) 2 ÷ 2	(34) 18 ÷ 2	(35) 10 ÷ 2
(36) 8 ÷ 2	(37) 14 ÷ 2	(38) 0 ÷ 2	(39) 14 ÷ 2	(40) 20 ÷ 2

7

Divide by 2

Time: __________

Score: _____/ 40

(1) 16 ÷ 2	(2) 12 ÷ 2	(3) 8 ÷ 2	(4) 4 ÷ 2	(5) 22 ÷ 2
(6) 10 ÷ 2	(7) 18 ÷ 2	(8) 0 ÷ 2	(9) 24 ÷ 2	(10) 6 ÷ 2
(11) 2 ÷ 2	(12) 6 ÷ 2	(13) 8 ÷ 2	(14) 2 ÷ 2	(15) 20 ÷ 2
(16) 0 ÷ 2	(17) 14 ÷ 2	(18) 22 ÷ 2	(19) 4 ÷ 2	(20) 16 ÷ 2
(21) 18 ÷ 2	(22) 24 ÷ 2	(23) 10 ÷ 2	(24) 12 ÷ 2	(25) 0 ÷ 2
(26) 4 ÷ 2	(27) 16 ÷ 2	(28) 8 ÷ 2	(29) 24 ÷ 2	(30) 22 ÷ 2
(31) 18 ÷ 2	(32) 12 ÷ 2	(33) 2 ÷ 2	(34) 20 ÷ 2	(35) 6 ÷ 2
(36) 14 ÷ 2	(37) 10 ÷ 2	(38) 12 ÷ 2	(39) 8 ÷ 2	(40) 14 ÷ 2

8

Divide by 2

Time: ________

Score: _____/ 40

① 4 ÷ 2	② 24 ÷ 2	③ 18 ÷ 2	④ 6 ÷ 2	⑤ 0 ÷ 2
⑥ 16 ÷ 2	⑦ 2 ÷ 2	⑧ 22 ÷ 2	⑨ 10 ÷ 2	⑩ 20 ÷ 2
⑪ 22 ÷ 2	⑫ 20 ÷ 2	⑬ 16 ÷ 2	⑭ 8 ÷ 2	⑮ 10 ÷ 2
⑯ 18 ÷ 2	⑰ 6 ÷ 2	⑱ 12 ÷ 2	⑲ 14 ÷ 2	⑳ 24 ÷ 2
(21) 0 ÷ 2	(22) 4 ÷ 2	(23) 2 ÷ 2	(24) 10 ÷ 2	(25) 8 ÷ 2
(26) 6 ÷ 2	(27) 0 ÷ 2	(28) 20 ÷ 2	(29) 24 ÷ 2	(30) 22 ÷ 2
(31) 4 ÷ 2	(32) 12 ÷ 2	(33) 16 ÷ 2	(34) 2 ÷ 2	(35) 18 ÷ 2
(36) 14 ÷ 2	(37) 18 ÷ 2	(38) 0 ÷ 2	(39) 24 ÷ 2	(40) 10 ÷ 2

9

Divide by 3

Time: __________

Score: _____/ 40

(1) 21 ÷ 3	(2) 6 ÷ 3	(3) 15 ÷ 3	(4) 9 ÷ 3	(5) 24 ÷ 3
(6) 0 ÷ 3	(7) 3 ÷ 3	(8) 12 ÷ 3	(9) 27 ÷ 3	(10) 18 ÷ 3
(11) 15 ÷ 3	(12) 9 ÷ 3	(13) 6 ÷ 3	(14) 0 ÷ 3	(15) 21 ÷ 3
(16) 27 ÷ 3	(17) 24 ÷ 3	(18) 18 ÷ 3	(19) 12 ÷ 3	(20) 3 ÷ 3
(21) 0 ÷ 3	(22) 9 ÷ 3	(23) 24 ÷ 3	(24) 12 ÷ 3	(25) 15 ÷ 3
(26) 27 ÷ 3	(27) 18 ÷ 3	(28) 3 ÷ 3	(29) 21 ÷ 3	(30) 6 ÷ 3
(31) 3 ÷ 3	(32) 18 ÷ 3	(33) 15 ÷ 3	(34) 0 ÷ 3	(35) 27 ÷ 3
(36) 6 ÷ 3	(37) 24 ÷ 3	(38) 21 ÷ 3	(39) 9 ÷ 3	(40) 12 ÷ 3

10

Divide by 3

Time: ____________

Score: ______/ 40

(1) 21 ÷ 3	(2) 12 ÷ 3	(3) 15 ÷ 3	(4) 24 ÷ 3	(5) 6 ÷ 3
(6) 3 ÷ 3	(7) 27 ÷ 3	(8) 18 ÷ 3	(9) 9 ÷ 3	(10) 0 ÷ 3
(11) 21 ÷ 3	(12) 15 ÷ 3	(13) 27 ÷ 3	(14) 9 ÷ 3	(15) 3 ÷ 3
(16) 12 ÷ 3	(17) 6 ÷ 3	(18) 24 ÷ 3	(19) 0 ÷ 3	(20) 18 ÷ 3
(21) 3 ÷ 3	(22) 21 ÷ 3	(23) 0 ÷ 3	(24) 18 ÷ 3	(25) 9 ÷ 3
(26) 12 ÷ 3	(27) 27 ÷ 3	(28) 6 ÷ 3	(29) 24 ÷ 3	(30) 15 ÷ 3
(31) 9 ÷ 3	(32) 3 ÷ 3	(33) 24 ÷ 3	(34) 21 ÷ 3	(35) 12 ÷ 3
(36) 15 ÷ 3	(37) 27 ÷ 3	(38) 0 ÷ 3	(39) 6 ÷ 3	(40) 18 ÷ 3

11

Divide by 3

Time: ____________

Score: ______/ 40

(1) 6 ÷ 3	(2) 27 ÷ 3	(3) 9 ÷ 3	(4) 21 ÷ 3	(5) 15 ÷ 3
(6) 18 ÷ 3	(7) 0 ÷ 3	(8) 12 ÷ 3	(9) 24 ÷ 3	(10) 3 ÷ 3
(11) 15 ÷ 3	(12) 3 ÷ 3	(13) 24 ÷ 3	(14) 27 ÷ 3	(15) 21 ÷ 3
(16) 0 ÷ 3	(17) 6 ÷ 3	(18) 9 ÷ 3	(19) 18 ÷ 3	(20) 12 ÷ 3
(21) 15 ÷ 3	(22) 12 ÷ 3	(23) 21 ÷ 3	(24) 9 ÷ 3	(25) 0 ÷ 3
(26) 6 ÷ 3	(27) 18 ÷ 3	(28) 3 ÷ 3	(29) 27 ÷ 3	(30) 24 ÷ 3
(31) 9 ÷ 3	(32) 27 ÷ 3	(33) 21 ÷ 3	(34) 15 ÷ 3	(35) 3 ÷ 3
(36) 0 ÷ 3	(37) 12 ÷ 3	(38) 6 ÷ 3	(39) 18 ÷ 3	(40) 24 ÷ 3

12

Divide by 3

Time: ___________

Score: _____/ 40

(1) 21 ÷ 3	(2) 27 ÷ 3	(3) 6 ÷ 3	(4) 9 ÷ 3	(5) 24 ÷ 3
(6) 15 ÷ 3	(7) 3 ÷ 3	(8) 18 ÷ 3	(9) 0 ÷ 3	(10) 12 ÷ 3
(11) 21 ÷ 3	(12) 12 ÷ 3	(13) 24 ÷ 3	(14) 9 ÷ 3	(15) 0 ÷ 3
(16) 15 ÷ 3	(17) 6 ÷ 3	(18) 3 ÷ 3	(19) 27 ÷ 3	(20) 18 ÷ 3
(21) 0 ÷ 3	(22) 12 ÷ 3	(23) 27 ÷ 3	(24) 6 ÷ 3	(25) 18 ÷ 3
(26) 21 ÷ 3	(27) 15 ÷ 3	(28) 24 ÷ 3	(29) 3 ÷ 3	(30) 9 ÷ 3
(31) 24 ÷ 3	(32) 9 ÷ 3	(33) 27 ÷ 3	(34) 18 ÷ 3	(35) 12 ÷ 3
(36) 6 ÷ 3	(37) 21 ÷ 3	(38) 3 ÷ 3	(39) 0 ÷ 3	(40) 15 ÷ 3

13

Divide by 1-3

Time: __________

Score: _____/ 40

(1) $2 \div 2$	(2) $6 \div 3$	(3) $0 \div 3$	(4) $12 \div 3$	(5) $6 \div 2$
(6) $9 \div 3$	(7) $24 \div 3$	(8) $3 \div 1$	(9) $6 \div 1$	(10) $4 \div 1$
(11) $0 \div 1$	(12) $4 \div 2$	(13) $14 \div 2$	(14) $8 \div 2$	(15) $21 \div 3$
(16) $7 \div 1$	(17) $18 \div 2$	(18) $1 \div 1$	(19) $0 \div 2$	(20) $3 \div 3$
(21) $15 \div 3$	(22) $16 \div 2$	(23) $12 \div 2$	(24) $8 \div 1$	(25) $5 \div 1$
(26) $27 \div 3$	(27) $2 \div 1$	(28) $18 \div 3$	(29) $10 \div 2$	(30) $9 \div 1$
(31) $21 \div 3$	(32) $12 \div 3$	(33) $8 \div 1$	(34) $15 \div 3$	(35) $9 \div 1$
(36) $1 \div 1$	(37) $0 \div 2$	(38) $4 \div 1$	(39) $4 \div 2$	(40) $12 \div 2$

14

Divide by 1-3

Time: ___________

Score: _____/ 40

(1) 6 ÷ 2	(2) 9 ÷ 3	(3) 27 ÷ 3	(4) 0 ÷ 1	(5) 7 ÷ 1
(6) 8 ÷ 2	(7) 5 ÷ 1	(8) 24 ÷ 3	(9) 3 ÷ 1	(10) 3 ÷ 3
(11) 10 ÷ 2	(12) 18 ÷ 3	(13) 0 ÷ 3	(14) 18 ÷ 2	(15) 2 ÷ 1
(16) 2 ÷ 2	(17) 14 ÷ 2	(18) 6 ÷ 1	(19) 6 ÷ 3	(20) 16 ÷ 2
(21) 2 ÷ 1	(22) 12 ÷ 2	(23) 12 ÷ 3	(24) 3 ÷ 1	(25) 4 ÷ 2
(26) 1 ÷ 1	(27) 7 ÷ 1	(28) 8 ÷ 1	(29) 9 ÷ 1	(30) 9 ÷ 3
(31) 6 ÷ 2	(32) 15 ÷ 3	(33) 10 ÷ 2	(34) 3 ÷ 3	(35) 24 ÷ 3
(36) 18 ÷ 2	(37) 4 ÷ 1	(38) 14 ÷ 2	(39) 6 ÷ 3	(40) 2 ÷ 2

15

Divide by 1-3

Time: ___________

Score: _____/ 40

① $5 \div 1$	② $0 \div 3$	③ $0 \div 2$	④ $0 \div 1$	⑤ $6 \div 1$
⑥ $21 \div 3$	⑦ $16 \div 2$	⑧ $27 \div 3$	⑨ $18 \div 3$	⑩ $8 \div 2$
⑪ $18 \div 3$	⑫ $12 \div 2$	⑬ $18 \div 2$	⑭ $15 \div 3$	⑮ $12 \div 3$
⑯ $14 \div 2$	⑰ $6 \div 3$	⑱ $4 \div 2$	⑲ $8 \div 1$	⑳ $6 \div 2$
21 $3 \div 3$	22 $27 \div 3$	23 $1 \div 1$	24 $2 \div 1$	25 $0 \div 1$
26 $8 \div 2$	27 $3 \div 1$	28 $21 \div 3$	29 $4 \div 1$	30 $10 \div 2$
31 $7 \div 1$	32 $6 \div 1$	33 $0 \div 3$	34 $9 \div 1$	35 $2 \div 2$
36 $24 \div 3$	37 $9 \div 3$	38 $16 \div 2$	39 $0 \div 2$	40 $5 \div 1$

16

Divide by 1-3

Time: ___________

Score: _____/ 40

① $9 \div 3$	② $4 \div 2$	③ $1 \div 1$	④ $3 \div 1$	⑤ $15 \div 3$
⑥ $12 \div 2$	⑦ $14 \div 2$	⑧ $24 \div 3$	⑨ $12 \div 3$	⑩ $27 \div 3$
⑪ $18 \div 3$	⑫ $3 \div 3$	⑬ $8 \div 2$	⑭ $9 \div 1$	⑮ $4 \div 1$
⑯ $16 \div 2$	⑰ $0 \div 2$	⑱ $2 \div 1$	⑲ $10 \div 2$	⑳ $8 \div 1$
(21) $0 \div 3$	(22) $0 \div 1$	(23) $5 \div 1$	(24) $6 \div 2$	(25) $6 \div 1$
(26) $6 \div 3$	(27) $7 \div 1$	(28) $18 \div 2$	(29) $21 \div 3$	(30) $2 \div 2$
(31) $8 \div 1$	(32) $4 \div 2$	(33) $5 \div 1$	(34) $6 \div 1$	(35) $18 \div 3$
(36) $3 \div 3$	(37) $7 \div 1$	(38) $2 \div 2$	(39) $9 \div 1$	(40) $12 \div 2$

17

Divide by 1-3

Time: ___________

Score: _____/ 40

(1) $12 \div 3$	(2) $9 \div 3$	(3) $27 \div 3$	(4) $14 \div 2$	(5) $2 \div 1$
(6) $4 \div 1$	(7) $3 \div 1$	(8) $18 \div 2$	(9) $21 \div 3$	(10) $15 \div 3$
(11) $0 \div 3$	(12) $0 \div 2$	(13) $6 \div 3$	(14) $10 \div 2$	(15) $16 \div 2$
(16) $0 \div 1$	(17) $6 \div 2$	(18) $8 \div 2$	(19) $24 \div 3$	(20) $1 \div 1$
(21) $21 \div 3$	(22) $12 \div 2$	(23) $0 \div 2$	(24) $27 \div 3$	(25) $6 \div 2$
(26) $0 \div 3$	(27) $9 \div 3$	(28) $2 \div 1$	(29) $3 \div 1$	(30) $8 \div 2$
(31) $9 \div 1$	(32) $1 \div 1$	(33) $10 \div 2$	(34) $8 \div 1$	(35) $0 \div 1$
(36) $2 \div 2$	(37) $12 \div 3$	(38) $15 \div 3$	(39) $18 \div 3$	(40) $3 \div 3$

18

Divide by 1-3

Time: ________

Score: _____/ 40

(1) $4 \div 2$	(2) $7 \div 1$	(3) $5 \div 1$	(4) $16 \div 2$	(5) $6 \div 3$
(6) $14 \div 2$	(7) $18 \div 2$	(8) $4 \div 1$	(9) $24 \div 3$	(10) $6 \div 1$
(11) $18 \div 3$	(12) $15 \div 3$	(13) $9 \div 1$	(14) $6 \div 1$	(15) $7 \div 1$
(16) $3 \div 3$	(17) $16 \div 2$	(18) $27 \div 3$	(19) $1 \div 1$	(20) $4 \div 1$
(21) $24 \div 3$	(22) $12 \div 2$	(23) $8 \div 1$	(24) $12 \div 3$	(25) $2 \div 2$
(26) $0 \div 3$	(27) $5 \div 1$	(28) $6 \div 2$	(29) $0 \div 2$	(30) $3 \div 1$
(31) $18 \div 2$	(32) $14 \div 2$	(33) $21 \div 3$	(34) $6 \div 3$	(35) $0 \div 1$
(36) $9 \div 3$	(37) $10 \div 2$	(38) $4 \div 2$	(39) $2 \div 1$	(40) $8 \div 2$

19

Divide by 1-3

Time: __________

Score: _____/ 40

① 9 ÷ 3	② 12 ÷ 2	③ 5 ÷ 1	④ 24 ÷ 3	⑤ 14 ÷ 2
⑥ 3 ÷ 3	⑦ 4 ÷ 2	⑧ 0 ÷ 2	⑨ 6 ÷ 3	⑩ 27 ÷ 3
⑪ 8 ÷ 2	⑫ 0 ÷ 3	⑬ 8 ÷ 1	⑭ 4 ÷ 1	⑮ 12 ÷ 3
⑯ 6 ÷ 1	⑰ 18 ÷ 2	⑱ 3 ÷ 1	⑲ 15 ÷ 3	⑳ 10 ÷ 2
(21) 2 ÷ 1	(22) 21 ÷ 3	(23) 0 ÷ 1	(24) 18 ÷ 3	(25) 6 ÷ 2
(26) 16 ÷ 2	(27) 7 ÷ 1	(28) 2 ÷ 2	(29) 1 ÷ 1	(30) 9 ÷ 1
(31) 21 ÷ 3	(32) 8 ÷ 2	(33) 6 ÷ 3	(34) 0 ÷ 2	(35) 9 ÷ 1
(36) 18 ÷ 2	(37) 16 ÷ 2	(38) 9 ÷ 3	(39) 0 ÷ 3	(40) 6 ÷ 1

20

Divide by 1-3

Time: ____________

Score: ______/ 40

1. 10 ÷ 2
2. 15 ÷ 3
3. 4 ÷ 2
4. 12 ÷ 2
5. 24 ÷ 3
6. 18 ÷ 3
7. 4 ÷ 1
8. 2 ÷ 2
9. 8 ÷ 1
10. 14 ÷ 2
11. 2 ÷ 1
12. 0 ÷ 1
13. 6 ÷ 2
14. 3 ÷ 1
15. 1 ÷ 1
16. 3 ÷ 3
17. 12 ÷ 3
18. 5 ÷ 1
19. 7 ÷ 1
20. 27 ÷ 3
21. 5 ÷ 1
22. 24 ÷ 3
23. 6 ÷ 3
24. 8 ÷ 2
25. 15 ÷ 3
26. 0 ÷ 1
27. 18 ÷ 2
28. 2 ÷ 2
29. 3 ÷ 1
30. 0 ÷ 3
31. 8 ÷ 1
32. 0 ÷ 2
33. 2 ÷ 1
34. 6 ÷ 2
35. 9 ÷ 1
36. 6 ÷ 1
37. 12 ÷ 2
38. 7 ÷ 1
39. 12 ÷ 3
40. 9 ÷ 3

21

Divide by 4

Time: ________

Score: _____/ 40

(1) $12 \div 4$	(2) $8 \div 4$	(3) $28 \div 4$	(4) $40 \div 4$	(5) $44 \div 4$
(6) $36 \div 4$	(7) $20 \div 4$	(8) $24 \div 4$	(9) $48 \div 4$	(10) $0 \div 4$
(11) $4 \div 4$	(12) $16 \div 4$	(13) $32 \div 4$	(14) $16 \div 4$	(15) $28 \div 4$
(16) $44 \div 4$	(17) $40 \div 4$	(18) $20 \div 4$	(19) $12 \div 4$	(20) $48 \div 4$
(21) $4 \div 4$	(22) $8 \div 4$	(23) $36 \div 4$	(24) $24 \div 4$	(25) $32 \div 4$
(26) $0 \div 4$	(27) $48 \div 4$	(28) $36 \div 4$	(29) $0 \div 4$	(30) $40 \div 4$
(31) $12 \div 4$	(32) $20 \div 4$	(33) $28 \div 4$	(34) $44 \div 4$	(35) $4 \div 4$
(36) $24 \div 4$	(37) $16 \div 4$	(38) $8 \div 4$	(39) $32 \div 4$	(40) $8 \div 4$

22

Divide by 4

Time: ________

Score: _____/ 40

(1) 32 ÷ 4	(2) 48 ÷ 4	(3) 4 ÷ 4	(4) 12 ÷ 4	(5) 24 ÷ 4
(6) 20 ÷ 4	(7) 16 ÷ 4	(8) 28 ÷ 4	(9) 40 ÷ 4	(10) 36 ÷ 4
(11) 44 ÷ 4	(12) 0 ÷ 4	(13) 36 ÷ 4	(14) 32 ÷ 4	(15) 28 ÷ 4
(16) 8 ÷ 4	(17) 44 ÷ 4	(18) 24 ÷ 4	(19) 0 ÷ 4	(20) 20 ÷ 4
(21) 48 ÷ 4	(22) 40 ÷ 4	(23) 12 ÷ 4	(24) 16 ÷ 4	(25) 4 ÷ 4
(26) 48 ÷ 4	(27) 8 ÷ 4	(28) 32 ÷ 4	(29) 12 ÷ 4	(30) 24 ÷ 4
(31) 28 ÷ 4	(32) 36 ÷ 4	(33) 0 ÷ 4	(34) 16 ÷ 4	(35) 44 ÷ 4
(36) 20 ÷ 4	(37) 4 ÷ 4	(38) 40 ÷ 4	(39) 20 ÷ 4	(40) 8 ÷ 4

23

Divide by 4

Time: ____________

Score: ______/ 40

(1) $0 \div 4$	(2) $28 \div 4$	(3) $40 \div 4$	(4) $4 \div 4$	(5) $36 \div 4$
(6) $24 \div 4$	(7) $44 \div 4$	(8) $48 \div 4$	(9) $16 \div 4$	(10) $32 \div 4$
(11) $12 \div 4$	(12) $28 \div 4$	(13) $8 \div 4$	(14) $24 \div 4$	(15) $36 \div 4$
(16) $16 \div 4$	(17) $44 \div 4$	(18) $48 \div 4$	(19) $40 \div 4$	(20) $20 \div 4$
(21) $32 \div 4$	(22) $0 \div 4$	(23) $4 \div 4$	(24) $12 \div 4$	(25) $24 \div 4$
(26) $28 \div 4$	(27) $36 \div 4$	(28) $40 \div 4$	(29) $20 \div 4$	(30) $16 \div 4$
(31) $12 \div 4$	(32) $48 \div 4$	(33) $44 \div 4$	(34) $32 \div 4$	(35) $0 \div 4$
(36) $4 \div 4$	(37) $8 \div 4$	(38) $16 \div 4$	(39) $0 \div 4$	(40) $36 \div 4$

24

Divide by 4

Time: ________

Score: _____/ 40

① $32 \div 4$	② $24 \div 4$	③ $28 \div 4$	④ $20 \div 4$	⑤ $48 \div 4$
⑥ $44 \div 4$	⑦ $4 \div 4$	⑧ $12 \div 4$	⑨ $8 \div 4$	⑩ $40 \div 4$
⑪ $4 \div 4$	⑫ $12 \div 4$	⑬ $40 \div 4$	⑭ $44 \div 4$	⑮ $0 \div 4$
⑯ $36 \div 4$	⑰ $24 \div 4$	⑱ $8 \div 4$	⑲ $48 \div 4$	⑳ $32 \div 4$
(21) $20 \div 4$	(22) $28 \div 4$	(23) $16 \div 4$	(24) $40 \div 4$	(25) $4 \div 4$
(26) $20 \div 4$	(27) $24 \div 4$	(28) $0 \div 4$	(29) $16 \div 4$	(30) $36 \div 4$
(31) $44 \div 4$	(32) $28 \div 4$	(33) $12 \div 4$	(34) $8 \div 4$	(35) $48 \div 4$
(36) $32 \div 4$	(37) $28 \div 4$	(38) $32 \div 4$	(39) $20 \div 4$	(40) $36 \div 4$

25

Divide by 5

Time: __________

Score: _____/ 40

(1) $5 \div 5$	(2) $30 \div 5$	(3) $25 \div 5$	(4) $45 \div 5$	(5) $0 \div 5$
(6) $50 \div 5$	(7) $20 \div 5$	(8) $60 \div 5$	(9) $15 \div 5$	(10) $55 \div 5$
(11) $40 \div 5$	(12) $10 \div 5$	(13) $35 \div 5$	(14) $55 \div 5$	(15) $15 \div 5$
(16) $25 \div 5$	(17) $20 \div 5$	(18) $30 \div 5$	(19) $50 \div 5$	(20) $40 \div 5$
(21) $35 \div 5$	(22) $60 \div 5$	(23) $5 \div 5$	(24) $10 \div 5$	(25) $45 \div 5$
(26) $0 \div 5$	(27) $30 \div 5$	(28) $50 \div 5$	(29) $5 \div 5$	(30) $15 \div 5$
(31) $40 \div 5$	(32) $35 \div 5$	(33) $55 \div 5$	(34) $10 \div 5$	(35) $45 \div 5$
(36) $25 \div 5$	(37) $0 \div 5$	(38) $60 \div 5$	(39) $20 \div 5$	(40) $0 \div 5$

26

Divide by 5

Time: ____________

Score: ______/ 40

(1) 50 ÷ 5	(2) 10 ÷ 5	(3) 30 ÷ 5	(4) 60 ÷ 5	(5) 55 ÷ 5
(6) 15 ÷ 5	(7) 5 ÷ 5	(8) 20 ÷ 5	(9) 35 ÷ 5	(10) 45 ÷ 5
(11) 40 ÷ 5	(12) 25 ÷ 5	(13) 45 ÷ 5	(14) 50 ÷ 5	(15) 10 ÷ 5
(16) 35 ÷ 5	(17) 5 ÷ 5	(18) 15 ÷ 5	(19) 25 ÷ 5	(20) 20 ÷ 5
(21) 30 ÷ 5	(22) 0 ÷ 5	(23) 40 ÷ 5	(24) 55 ÷ 5	(25) 60 ÷ 5
(26) 10 ÷ 5	(27) 20 ÷ 5	(28) 60 ÷ 5	(29) 5 ÷ 5	(30) 30 ÷ 5
(31) 0 ÷ 5	(32) 55 ÷ 5	(33) 15 ÷ 5	(34) 45 ÷ 5	(35) 35 ÷ 5
(36) 40 ÷ 5	(37) 25 ÷ 5	(38) 50 ÷ 5	(39) 10 ÷ 5	(40) 45 ÷ 5

27

Divide by 5

Time: ___________

Score: _____/ 40

(1) 25 ÷ 5	(2) 35 ÷ 5	(3) 55 ÷ 5	(4) 0 ÷ 5	(5) 30 ÷ 5
(6) 50 ÷ 5	(7) 40 ÷ 5	(8) 20 ÷ 5	(9) 60 ÷ 5	(10) 15 ÷ 5
(11) 5 ÷ 5	(12) 0 ÷ 5	(13) 45 ÷ 5	(14) 5 ÷ 5	(15) 60 ÷ 5
(16) 10 ÷ 5	(17) 25 ÷ 5	(18) 35 ÷ 5	(19) 55 ÷ 5	(20) 40 ÷ 5
(21) 50 ÷ 5	(22) 30 ÷ 5	(23) 15 ÷ 5	(24) 20 ÷ 5	(25) 25 ÷ 5
(26) 10 ÷ 5	(27) 45 ÷ 5	(28) 15 ÷ 5	(29) 35 ÷ 5	(30) 20 ÷ 5
(31) 60 ÷ 5	(32) 5 ÷ 5	(33) 50 ÷ 5	(34) 55 ÷ 5	(35) 30 ÷ 5
(36) 40 ÷ 5	(37) 0 ÷ 5	(38) 25 ÷ 5	(39) 0 ÷ 5	(40) 15 ÷ 5

28

Divide by 5

Time: ____________

Score: _____/ 40

(1) 20 ÷ 5	(2) 5 ÷ 5	(3) 55 ÷ 5	(4) 60 ÷ 5	(5) 50 ÷ 5
(6) 10 ÷ 5	(7) 35 ÷ 5	(8) 45 ÷ 5	(9) 30 ÷ 5	(10) 40 ÷ 5
(11) 50 ÷ 5	(12) 20 ÷ 5	(13) 25 ÷ 5	(14) 15 ÷ 5	(15) 30 ÷ 5
(16) 55 ÷ 5	(17) 5 ÷ 5	(18) 0 ÷ 5	(19) 10 ÷ 5	(20) 60 ÷ 5
(21) 35 ÷ 5	(22) 40 ÷ 5	(23) 45 ÷ 5	(24) 35 ÷ 5	(25) 60 ÷ 5
(26) 20 ÷ 5	(27) 50 ÷ 5	(28) 0 ÷ 5	(29) 5 ÷ 5	(30) 10 ÷ 5
(31) 15 ÷ 5	(32) 45 ÷ 5	(33) 40 ÷ 5	(34) 30 ÷ 5	(35) 55 ÷ 5
(36) 25 ÷ 5	(37) 30 ÷ 5	(38) 35 ÷ 5	(39) 45 ÷ 5	(40) 40 ÷ 5

29

Divide by 6

Time: ____________

Score: _____/ 40

(1) $42 \div 6$	(2) $0 \div 6$	(3) $54 \div 6$	(4) $66 \div 6$	(5) $72 \div 6$
(6) $60 \div 6$	(7) $30 \div 6$	(8) $24 \div 6$	(9) $18 \div 6$	(10) $48 \div 6$
(11) $12 \div 6$	(12) $6 \div 6$	(13) $36 \div 6$	(14) $66 \div 6$	(15) $36 \div 6$
(16) $30 \div 6$	(17) $60 \div 6$	(18) $12 \div 6$	(19) $54 \div 6$	(20) $18 \div 6$
(21) $6 \div 6$	(22) $0 \div 6$	(23) $24 \div 6$	(24) $42 \div 6$	(25) $48 \div 6$
(26) $72 \div 6$	(27) $42 \div 6$	(28) $72 \div 6$	(29) $48 \div 6$	(30) $60 \div 6$
(31) $18 \div 6$	(32) $0 \div 6$	(33) $6 \div 6$	(34) $24 \div 6$	(35) $12 \div 6$
(36) $30 \div 6$	(37) $36 \div 6$	(38) $66 \div 6$	(39) $54 \div 6$	(40) $6 \div 6$

30

Divide by 6

Time: ____________

Score: ______/ 40

(1) 12 ÷ 6	(2) 30 ÷ 6	(3) 72 ÷ 6	(4) 60 ÷ 6	(5) 36 ÷ 6
(6) 0 ÷ 6	(7) 24 ÷ 6	(8) 66 ÷ 6	(9) 48 ÷ 6	(10) 42 ÷ 6
(11) 18 ÷ 6	(12) 54 ÷ 6	(13) 18 ÷ 6	(14) 42 ÷ 6	(15) 24 ÷ 6
(16) 36 ÷ 6	(17) 54 ÷ 6	(18) 60 ÷ 6	(19) 12 ÷ 6	(20) 72 ÷ 6
(21) 48 ÷ 6	(22) 30 ÷ 6	(23) 6 ÷ 6	(24) 66 ÷ 6	(25) 0 ÷ 6
(26) 36 ÷ 6	(27) 60 ÷ 6	(28) 48 ÷ 6	(29) 54 ÷ 6	(30) 24 ÷ 6
(31) 72 ÷ 6	(32) 18 ÷ 6	(33) 0 ÷ 6	(34) 42 ÷ 6	(35) 30 ÷ 6
(36) 12 ÷ 6	(37) 6 ÷ 6	(38) 66 ÷ 6	(39) 48 ÷ 6	(40) 66 ÷ 6

31

Divide by 6

Time: ________

Score: _____/ 40

① 36 ÷ 6	② 0 ÷ 6	③ 24 ÷ 6	④ 42 ÷ 6	⑤ 6 ÷ 6
⑥ 18 ÷ 6	⑦ 60 ÷ 6	⑧ 72 ÷ 6	⑨ 54 ÷ 6	⑩ 12 ÷ 6
⑪ 30 ÷ 6	⑫ 66 ÷ 6	⑬ 6 ÷ 6	⑭ 30 ÷ 6	⑮ 18 ÷ 6
⑯ 0 ÷ 6	⑰ 54 ÷ 6	⑱ 42 ÷ 6	⑲ 12 ÷ 6	⑳ 36 ÷ 6
(21) 60 ÷ 6	(22) 72 ÷ 6	(23) 24 ÷ 6	(24) 48 ÷ 6	(25) 30 ÷ 6
(26) 66 ÷ 6	(27) 48 ÷ 6	(28) 0 ÷ 6	(29) 24 ÷ 6	(30) 72 ÷ 6
(31) 12 ÷ 6	(32) 42 ÷ 6	(33) 18 ÷ 6	(34) 36 ÷ 6	(35) 6 ÷ 6
(36) 54 ÷ 6	(37) 60 ÷ 6	(38) 30 ÷ 6	(39) 60 ÷ 6	(40) 12 ÷ 6

32

Divide by 6

Time: ___________

Score: _____/ 40

(1) 36 ÷ 6	(2) 72 ÷ 6	(3) 42 ÷ 6	(4) 18 ÷ 6	(5) 48 ÷ 6
(6) 24 ÷ 6	(7) 54 ÷ 6	(8) 0 ÷ 6	(9) 6 ÷ 6	(10) 66 ÷ 6
(11) 42 ÷ 6	(12) 48 ÷ 6	(13) 72 ÷ 6	(14) 30 ÷ 6	(15) 18 ÷ 6
(16) 6 ÷ 6	(17) 66 ÷ 6	(18) 0 ÷ 6	(19) 24 ÷ 6	(20) 54 ÷ 6
(21) 36 ÷ 6	(22) 12 ÷ 6	(23) 60 ÷ 6	(24) 0 ÷ 6	(25) 66 ÷ 6
(26) 6 ÷ 6	(27) 12 ÷ 6	(28) 48 ÷ 6	(29) 36 ÷ 6	(30) 54 ÷ 6
(31) 42 ÷ 6	(32) 30 ÷ 6	(33) 72 ÷ 6	(34) 24 ÷ 6	(35) 18 ÷ 6
(36) 60 ÷ 6	(37) 6 ÷ 6	(38) 54 ÷ 6	(39) 36 ÷ 6	(40) 0 ÷ 6

33

Divide by 4-6

Time: ___________

Score: _____/ 40

1. 12 ÷ 6
2. 50 ÷ 5
3. 44 ÷ 4
4. 4 ÷ 4
5. 12 ÷ 4
6. 5 ÷ 5
7. 40 ÷ 5
8. 55 ÷ 5
9. 66 ÷ 6
10. 28 ÷ 4
11. 24 ÷ 4
12. 48 ÷ 6
13. 10 ÷ 5
14. 36 ÷ 4
15. 45 ÷ 5
16. 16 ÷ 4
17. 60 ÷ 6
18. 25 ÷ 5
19. 30 ÷ 6
20. 8 ÷ 4
21. 6 ÷ 6
22. 40 ÷ 4
23. 54 ÷ 6
24. 35 ÷ 5
25. 60 ÷ 5
26. 24 ÷ 6
27. 36 ÷ 6
28. 0 ÷ 4
29. 0 ÷ 6
30. 72 ÷ 6
31. 0 ÷ 5
32. 30 ÷ 5
33. 20 ÷ 5
34. 15 ÷ 5
35. 18 ÷ 6
36. 32 ÷ 4
37. 42 ÷ 6
38. 48 ÷ 4
39. 20 ÷ 4
40. 45 ÷ 5

34

Divide by 4-6

Time: ____________

Score: ______/ 40

① $20 \div 5$	② $6 \div 6$	③ $36 \div 6$	④ $66 \div 6$	⑤ $36 \div 4$
⑥ $24 \div 4$	⑦ $25 \div 5$	⑧ $72 \div 6$	⑨ $54 \div 6$	⑩ $55 \div 5$
⑪ $40 \div 4$	⑫ $44 \div 4$	⑬ $60 \div 5$	⑭ $30 \div 6$	⑮ $28 \div 4$
⑯ $48 \div 4$	⑰ $0 \div 5$	⑱ $12 \div 6$	⑲ $5 \div 5$	⑳ $16 \div 4$
(21) $4 \div 4$	(22) $0 \div 4$	(23) $8 \div 4$	(24) $40 \div 5$	(25) $30 \div 5$
(26) $60 \div 6$	(27) $12 \div 4$	(28) $24 \div 6$	(29) $42 \div 6$	(30) $18 \div 6$
(31) $15 \div 5$	(32) $35 \div 5$	(33) $48 \div 6$	(34) $10 \div 5$	(35) $20 \div 4$
(36) $50 \div 5$	(37) $32 \div 4$	(38) $0 \div 6$	(39) $48 \div 4$	(40) $6 \div 6$

35

Divide by 4-6

Time: __________

Score: _____/ 40

(1) 32 ÷ 4	(2) 8 ÷ 4	(3) 15 ÷ 5	(4) 36 ÷ 6	(5) 4 ÷ 4
(6) 0 ÷ 6	(7) 12 ÷ 6	(8) 60 ÷ 6	(9) 30 ÷ 6	(10) 10 ÷ 5
(11) 72 ÷ 6	(12) 48 ÷ 6	(13) 0 ÷ 5	(14) 44 ÷ 4	(15) 40 ÷ 5
(16) 30 ÷ 5	(17) 35 ÷ 5	(18) 24 ÷ 4	(19) 66 ÷ 6	(20) 18 ÷ 6
(21) 5 ÷ 5	(22) 50 ÷ 5	(23) 25 ÷ 5	(24) 55 ÷ 5	(25) 28 ÷ 4
(26) 42 ÷ 6	(27) 20 ÷ 4	(28) 36 ÷ 4	(29) 0 ÷ 4	(30) 45 ÷ 5
(31) 40 ÷ 4	(32) 12 ÷ 4	(33) 24 ÷ 6	(34) 54 ÷ 6	(35) 20 ÷ 5
(36) 60 ÷ 5	(37) 16 ÷ 4	(38) 25 ÷ 5	(39) 40 ÷ 4	(40) 18 ÷ 6

36

Divide by 4-6

Time: __________

Score: _____/ 40

1. $42 \div 6$
2. $12 \div 6$
3. $0 \div 6$
4. $20 \div 4$
5. $35 \div 5$
6. $15 \div 5$
7. $10 \div 5$
8. $60 \div 6$
9. $60 \div 5$
10. $0 \div 4$
11. $0 \div 5$
12. $45 \div 5$
13. $24 \div 4$
14. $6 \div 6$
15. $30 \div 6$
16. $24 \div 6$
17. $48 \div 4$
18. $66 \div 6$
19. $32 \div 4$
20. $28 \div 4$
21. $12 \div 4$
22. $54 \div 6$
23. $16 \div 4$
24. $72 \div 6$
25. $30 \div 5$
26. $4 \div 4$
27. $50 \div 5$
28. $5 \div 5$
29. $36 \div 6$
30. $36 \div 4$
31. $40 \div 5$
32. $20 \div 5$
33. $44 \div 4$
34. $55 \div 5$
35. $8 \div 4$
36. $48 \div 6$
37. $30 \div 5$
38. $48 \div 4$
39. $10 \div 5$
40. $5 \div 5$

37

Divide by 4-6

Time: ___________

Score: _____/ 40

(1) 18 ÷ 6	(2) 44 ÷ 4	(3) 28 ÷ 4	(4) 12 ÷ 6	(5) 4 ÷ 4
(6) 48 ÷ 6	(7) 12 ÷ 4	(8) 40 ÷ 4	(9) 60 ÷ 6	(10) 36 ÷ 4
(11) 20 ÷ 4	(12) 55 ÷ 5	(13) 0 ÷ 6	(14) 0 ÷ 5	(15) 50 ÷ 5
(16) 72 ÷ 6	(17) 40 ÷ 5	(18) 60 ÷ 5	(19) 25 ÷ 5	(20) 24 ÷ 6
(21) 54 ÷ 6	(22) 45 ÷ 5	(23) 20 ÷ 5	(24) 66 ÷ 6	(25) 0 ÷ 4
(26) 24 ÷ 4	(27) 15 ÷ 5	(28) 16 ÷ 4	(29) 6 ÷ 6	(30) 8 ÷ 4
(31) 30 ÷ 6	(32) 32 ÷ 4	(33) 35 ÷ 5	(34) 42 ÷ 6	(35) 36 ÷ 6
(36) 12 ÷ 4	(37) 15 ÷ 5	(38) 8 ÷ 4	(39) 60 ÷ 6	(40) 25 ÷ 5

38

Divide by 4-6

Time: ___________

Score: _____/ 40

① 6 ÷ 6	② 0 ÷ 6	③ 20 ÷ 5	④ 36 ÷ 6	⑤ 48 ÷ 4
⑥ 42 ÷ 6	⑦ 30 ÷ 5	⑧ 60 ÷ 5	⑨ 5 ÷ 5	⑩ 4 ÷ 4
⑪ 44 ÷ 4	⑫ 45 ÷ 5	⑬ 36 ÷ 4	⑭ 16 ÷ 4	⑮ 12 ÷ 6
⑯ 24 ÷ 4	⑰ 0 ÷ 5	⑱ 72 ÷ 6	⑲ 0 ÷ 4	⑳ 40 ÷ 4
(21) 35 ÷ 5	(22) 24 ÷ 6	(23) 40 ÷ 5	(24) 55 ÷ 5	(25) 28 ÷ 4
(26) 66 ÷ 6	(27) 32 ÷ 4	(28) 18 ÷ 6	(29) 10 ÷ 5	(30) 48 ÷ 6
(31) 20 ÷ 4	(32) 54 ÷ 6	(33) 30 ÷ 6	(34) 50 ÷ 5	(35) 20 ÷ 5
(36) 4 ÷ 4	(37) 25 ÷ 5	(38) 20 ÷ 4	(39) 30 ÷ 5	(40) 36 ÷ 6

39

Divide by 4-6

Time: ____________

Score: ______/ 40

(1) 48 ÷ 4	(2) 36 ÷ 4	(3) 12 ÷ 6	(4) 48 ÷ 6	(5) 6 ÷ 6
(6) 10 ÷ 5	(7) 0 ÷ 6	(8) 50 ÷ 5	(9) 30 ÷ 6	(10) 0 ÷ 5
(11) 72 ÷ 6	(12) 60 ÷ 6	(13) 15 ÷ 5	(14) 18 ÷ 6	(15) 40 ÷ 4
(16) 54 ÷ 6	(17) 28 ÷ 4	(18) 55 ÷ 5	(19) 5 ÷ 5	(20) 32 ÷ 4
(21) 16 ÷ 4	(22) 24 ÷ 6	(23) 60 ÷ 5	(24) 44 ÷ 4	(25) 12 ÷ 4
(26) 66 ÷ 6	(27) 24 ÷ 4	(28) 40 ÷ 5	(29) 8 ÷ 4	(30) 42 ÷ 6
(31) 45 ÷ 5	(32) 0 ÷ 4	(33) 35 ÷ 5	(34) 28 ÷ 4	(35) 12 ÷ 6
(36) 24 ÷ 4	(37) 5 ÷ 5	(38) 54 ÷ 6	(39) 20 ÷ 4	(40) 0 ÷ 4

40

Divide by 4-6

Time: __________

Score: _____/ 40

① 36 ÷ 6	② 30 ÷ 5	③ 45 ÷ 5	④ 40 ÷ 4	⑤ 8 ÷ 4
⑥ 16 ÷ 4	⑦ 24 ÷ 6	⑧ 25 ÷ 5	⑨ 12 ÷ 4	⑩ 32 ÷ 4
⑪ 40 ÷ 5	⑫ 20 ÷ 5	⑬ 55 ÷ 5	⑭ 35 ÷ 5	⑮ 50 ÷ 5
⑯ 30 ÷ 6	⑰ 48 ÷ 4	⑱ 0 ÷ 5	⑲ 60 ÷ 5	⑳ 15 ÷ 5
(21) 4 ÷ 4	(22) 0 ÷ 6	(23) 10 ÷ 5	(24) 6 ÷ 6	(25) 60 ÷ 6
(26) 44 ÷ 4	(27) 18 ÷ 6	(28) 72 ÷ 6	(29) 48 ÷ 6	(30) 42 ÷ 6
(31) 66 ÷ 6	(32) 36 ÷ 4	(33) 4 ÷ 4	(34) 10 ÷ 5	(35) 30 ÷ 6
(36) 0 ÷ 4	(37) 45 ÷ 5	(38) 54 ÷ 6	(39) 6 ÷ 6	(40) 42 ÷ 6

41

Divide by 7

Time: ________

Score: _____/ 40

(1) 49 ÷ 7	(2) 28 ÷ 7	(3) 84 ÷ 7	(4) 14 ÷ 7	(5) 42 ÷ 7
(6) 35 ÷ 7	(7) 63 ÷ 7	(8) 56 ÷ 7	(9) 0 ÷ 7	(10) 77 ÷ 7
(11) 7 ÷ 7	(12) 21 ÷ 7	(13) 70 ÷ 7	(14) 77 ÷ 7	(15) 70 ÷ 7
(16) 42 ÷ 7	(17) 21 ÷ 7	(18) 0 ÷ 7	(19) 56 ÷ 7	(20) 14 ÷ 7
(21) 35 ÷ 7	(22) 63 ÷ 7	(23) 28 ÷ 7	(24) 49 ÷ 7	(25) 7 ÷ 7
(26) 84 ÷ 7	(27) 35 ÷ 7	(28) 63 ÷ 7	(29) 84 ÷ 7	(30) 14 ÷ 7
(31) 0 ÷ 7	(32) 70 ÷ 7	(33) 77 ÷ 7	(34) 56 ÷ 7	(35) 42 ÷ 7
(36) 28 ÷ 7	(37) 7 ÷ 7	(38) 49 ÷ 7	(39) 21 ÷ 7	(40) 56 ÷ 7

42

Divide by 7

Time: ____________

Score: ______/ 40

(1) 84 ÷ 7	(2) 35 ÷ 7	(3) 42 ÷ 7	(4) 0 ÷ 7	(5) 63 ÷ 7
(6) 14 ÷ 7	(7) 7 ÷ 7	(8) 21 ÷ 7	(9) 77 ÷ 7	(10) 70 ÷ 7
(11) 28 ÷ 7	(12) 49 ÷ 7	(13) 63 ÷ 7	(14) 49 ÷ 7	(15) 77 ÷ 7
(16) 7 ÷ 7	(17) 0 ÷ 7	(18) 42 ÷ 7	(19) 35 ÷ 7	(20) 21 ÷ 7
(21) 14 ÷ 7	(22) 70 ÷ 7	(23) 56 ÷ 7	(24) 84 ÷ 7	(25) 28 ÷ 7
(26) 14 ÷ 7	(27) 70 ÷ 7	(28) 49 ÷ 7	(29) 7 ÷ 7	(30) 42 ÷ 7
(31) 63 ÷ 7	(32) 84 ÷ 7	(33) 0 ÷ 7	(34) 35 ÷ 7	(35) 21 ÷ 7
(36) 56 ÷ 7	(37) 28 ÷ 7	(38) 77 ÷ 7	(39) 49 ÷ 7	(40) 84 ÷ 7

43

Divide by 7

Time: ________

Score: _____/ 40

(1) 35 ÷ 7	(2) 42 ÷ 7	(3) 77 ÷ 7	(4) 28 ÷ 7	(5) 7 ÷ 7
(6) 14 ÷ 7	(7) 63 ÷ 7	(8) 70 ÷ 7	(9) 0 ÷ 7	(10) 21 ÷ 7
(11) 56 ÷ 7	(12) 77 ÷ 7	(13) 21 ÷ 7	(14) 70 ÷ 7	(15) 42 ÷ 7
(16) 56 ÷ 7	(17) 35 ÷ 7	(18) 84 ÷ 7	(19) 0 ÷ 7	(20) 14 ÷ 7
(21) 28 ÷ 7	(22) 49 ÷ 7	(23) 7 ÷ 7	(24) 63 ÷ 7	(25) 0 ÷ 7
(26) 63 ÷ 7	(27) 14 ÷ 7	(28) 28 ÷ 7	(29) 77 ÷ 7	(30) 7 ÷ 7
(31) 35 ÷ 7	(32) 70 ÷ 7	(33) 21 ÷ 7	(34) 56 ÷ 7	(35) 42 ÷ 7
(36) 49 ÷ 7	(37) 84 ÷ 7	(38) 14 ÷ 7	(39) 70 ÷ 7	(40) 84 ÷ 7

44

Divide by 7

Time: ___________

Score: _____/ 40

(1) 0 ÷ 7	(2) 49 ÷ 7	(3) 35 ÷ 7	(4) 56 ÷ 7	(5) 77 ÷ 7
(6) 21 ÷ 7	(7) 7 ÷ 7	(8) 63 ÷ 7	(9) 42 ÷ 7	(10) 28 ÷ 7
(11) 42 ÷ 7	(12) 84 ÷ 7	(13) 0 ÷ 7	(14) 35 ÷ 7	(15) 56 ÷ 7
(16) 70 ÷ 7	(17) 14 ÷ 7	(18) 49 ÷ 7	(19) 28 ÷ 7	(20) 7 ÷ 7
(21) 21 ÷ 7	(22) 77 ÷ 7	(23) 63 ÷ 7	(24) 0 ÷ 7	(25) 84 ÷ 7
(26) 77 ÷ 7	(27) 21 ÷ 7	(28) 14 ÷ 7	(29) 49 ÷ 7	(30) 35 ÷ 7
(31) 42 ÷ 7	(32) 28 ÷ 7	(33) 63 ÷ 7	(34) 56 ÷ 7	(35) 7 ÷ 7
(36) 70 ÷ 7	(37) 35 ÷ 7	(38) 77 ÷ 7	(39) 0 ÷ 7	(40) 70 ÷ 7

45

Divide by 8

Time: __________

Score: _____/ 40

① 16 ÷ 8	② 56 ÷ 8	③ 80 ÷ 8	④ 72 ÷ 8	⑤ 40 ÷ 8
⑥ 64 ÷ 8	⑦ 8 ÷ 8	⑧ 96 ÷ 8	⑨ 88 ÷ 8	⑩ 48 ÷ 8
⑪ 24 ÷ 8	⑫ 0 ÷ 8	⑬ 32 ÷ 8	⑭ 80 ÷ 8	⑮ 32 ÷ 8
⑯ 88 ÷ 8	⑰ 72 ÷ 8	⑱ 24 ÷ 8	⑲ 40 ÷ 8	⑳ 0 ÷ 8
(21) 64 ÷ 8	(22) 16 ÷ 8	(23) 48 ÷ 8	(24) 8 ÷ 8	(25) 96 ÷ 8
(26) 56 ÷ 8	(27) 88 ÷ 8	(28) 56 ÷ 8	(29) 80 ÷ 8	(30) 8 ÷ 8
(31) 40 ÷ 8	(32) 64 ÷ 8	(33) 24 ÷ 8	(34) 16 ÷ 8	(35) 96 ÷ 8
(36) 0 ÷ 8	(37) 72 ÷ 8	(38) 32 ÷ 8	(39) 48 ÷ 8	(40) 40 ÷ 8

46

Divide by 8

Time: ________

Score: _____/ 40

(1) $48 \div 8$	(2) $64 \div 8$	(3) $0 \div 8$	(4) $56 \div 8$	(5) $80 \div 8$
(6) $8 \div 8$	(7) $72 \div 8$	(8) $96 \div 8$	(9) $32 \div 8$	(10) $16 \div 8$
(11) $24 \div 8$	(12) $88 \div 8$	(13) $48 \div 8$	(14) $64 \div 8$	(15) $80 \div 8$
(16) $56 \div 8$	(17) $88 \div 8$	(18) $24 \div 8$	(19) $8 \div 8$	(20) $72 \div 8$
(21) $96 \div 8$	(22) $40 \div 8$	(23) $16 \div 8$	(24) $32 \div 8$	(25) $0 \div 8$
(26) $24 \div 8$	(27) $80 \div 8$	(28) $72 \div 8$	(29) $64 \div 8$	(30) $96 \div 8$
(31) $0 \div 8$	(32) $16 \div 8$	(33) $56 \div 8$	(34) $48 \div 8$	(35) $32 \div 8$
(36) $40 \div 8$	(37) $88 \div 8$	(38) $8 \div 8$	(39) $40 \div 8$	(40) $72 \div 8$

47

Divide by 8

Time: ___________

Score: _____/ 40

(1) $64 \div 8$	(2) $16 \div 8$	(3) $8 \div 8$	(4) $0 \div 8$	(5) $88 \div 8$
(6) $80 \div 8$	(7) $32 \div 8$	(8) $96 \div 8$	(9) $48 \div 8$	(10) $56 \div 8$
(11) $24 \div 8$	(12) $0 \div 8$	(13) $56 \div 8$	(14) $40 \div 8$	(15) $48 \div 8$
(16) $72 \div 8$	(17) $24 \div 8$	(18) $80 \div 8$	(19) $8 \div 8$	(20) $64 \div 8$
(21) $96 \div 8$	(22) $16 \div 8$	(23) $88 \div 8$	(24) $32 \div 8$	(25) $72 \div 8$
(26) $24 \div 8$	(27) $56 \div 8$	(28) $8 \div 8$	(29) $80 \div 8$	(30) $88 \div 8$
(31) $64 \div 8$	(32) $96 \div 8$	(33) $16 \div 8$	(34) $48 \div 8$	(35) $32 \div 8$
(36) $40 \div 8$	(37) $0 \div 8$	(38) $96 \div 8$	(39) $64 \div 8$	(40) $56 \div 8$

48

Divide by 8

Time: ________

Score: _____/ 40

(1) 88 ÷ 8	(2) 80 ÷ 8	(3) 32 ÷ 8	(4) 72 ÷ 8	(5) 40 ÷ 8
(6) 16 ÷ 8	(7) 0 ÷ 8	(8) 48 ÷ 8	(9) 8 ÷ 8	(10) 24 ÷ 8
(11) 96 ÷ 8	(12) 40 ÷ 8	(13) 88 ÷ 8	(14) 48 ÷ 8	(15) 64 ÷ 8
(16) 0 ÷ 8	(17) 72 ÷ 8	(18) 80 ÷ 8	(19) 56 ÷ 8	(20) 16 ÷ 8
(21) 32 ÷ 8	(22) 8 ÷ 8	(23) 24 ÷ 8	(24) 40 ÷ 8	(25) 24 ÷ 8
(26) 32 ÷ 8	(27) 16 ÷ 8	(28) 88 ÷ 8	(29) 8 ÷ 8	(30) 48 ÷ 8
(31) 96 ÷ 8	(32) 80 ÷ 8	(33) 64 ÷ 8	(34) 56 ÷ 8	(35) 72 ÷ 8
(36) 0 ÷ 8	(37) 96 ÷ 8	(38) 80 ÷ 8	(39) 8 ÷ 8	(40) 88 ÷ 8

49

Divide by 9

Time: ________

Score: _____/ 40

(1) $18 \div 9$	(2) $63 \div 9$	(3) $54 \div 9$	(4) $36 \div 9$	(5) $9 \div 9$
(6) $99 \div 9$	(7) $0 \div 9$	(8) $72 \div 9$	(9) $81 \div 9$	(10) $27 \div 9$
(11) $45 \div 9$	(12) $90 \div 9$	(13) $108 \div 9$	(14) $81 \div 9$	(15) $27 \div 9$
(16) $54 \div 9$	(17) $0 \div 9$	(18) $90 \div 9$	(19) $108 \div 9$	(20) $72 \div 9$
(21) $36 \div 9$	(22) $99 \div 9$	(23) $63 \div 9$	(24) $45 \div 9$	(25) $18 \div 9$
(26) $9 \div 9$	(27) $54 \div 9$	(28) $72 \div 9$	(29) $27 \div 9$	(30) $0 \div 9$
(31) $18 \div 9$	(32) $81 \div 9$	(33) $99 \div 9$	(34) $63 \div 9$	(35) $45 \div 9$
(36) $9 \div 9$	(37) $108 \div 9$	(38) $36 \div 9$	(39) $90 \div 9$	(40) $99 \div 9$

50

Divide by 9

Time: ___________

Score: _____/ 40

(1) 72 ÷ 9	(2) 54 ÷ 9	(3) 81 ÷ 9	(4) 27 ÷ 9	(5) 45 ÷ 9
(6) 108 ÷ 9	(7) 0 ÷ 9	(8) 9 ÷ 9	(9) 63 ÷ 9	(10) 36 ÷ 9
(11) 90 ÷ 9	(12) 18 ÷ 9	(13) 45 ÷ 9	(14) 18 ÷ 9	(15) 99 ÷ 9
(16) 108 ÷ 9	(17) 81 ÷ 9	(18) 90 ÷ 9	(19) 54 ÷ 9	(20) 36 ÷ 9
(21) 0 ÷ 9	(22) 9 ÷ 9	(23) 63 ÷ 9	(24) 27 ÷ 9	(25) 72 ÷ 9
(26) 18 ÷ 9	(27) 0 ÷ 9	(28) 72 ÷ 9	(29) 27 ÷ 9	(30) 108 ÷ 9
(31) 36 ÷ 9	(32) 90 ÷ 9	(33) 99 ÷ 9	(34) 45 ÷ 9	(35) 9 ÷ 9
(36) 81 ÷ 9	(37) 63 ÷ 9	(38) 54 ÷ 9	(39) 18 ÷ 9	(40) 36 ÷ 9

51

Divide by 9

Time: ________

Score: _____/ 40

(1) 99 ÷ 9	(2) 72 ÷ 9	(3) 63 ÷ 9	(4) 90 ÷ 9	(5) 45 ÷ 9
(6) 27 ÷ 9	(7) 81 ÷ 9	(8) 9 ÷ 9	(9) 0 ÷ 9	(10) 108 ÷ 9
(11) 54 ÷ 9	(12) 90 ÷ 9	(13) 54 ÷ 9	(14) 45 ÷ 9	(15) 0 ÷ 9
(16) 63 ÷ 9	(17) 36 ÷ 9	(18) 108 ÷ 9	(19) 27 ÷ 9	(20) 72 ÷ 9
(21) 9 ÷ 9	(22) 81 ÷ 9	(23) 18 ÷ 9	(24) 99 ÷ 9	(25) 9 ÷ 9
(26) 99 ÷ 9	(27) 45 ÷ 9	(28) 90 ÷ 9	(29) 0 ÷ 9	(30) 108 ÷ 9
(31) 18 ÷ 9	(32) 72 ÷ 9	(33) 27 ÷ 9	(34) 54 ÷ 9	(35) 36 ÷ 9
(36) 63 ÷ 9	(37) 81 ÷ 9	(38) 18 ÷ 9	(39) 63 ÷ 9	(40) 0 ÷ 9

52

Divide by 9

Time: ________

Score: _____/ 40

(1) 90 ÷ 9	(2) 72 ÷ 9	(3) 99 ÷ 9	(4) 45 ÷ 9	(5) 108 ÷ 9
(6) 36 ÷ 9	(7) 9 ÷ 9	(8) 81 ÷ 9	(9) 54 ÷ 9	(10) 27 ÷ 9
(11) 45 ÷ 9	(12) 72 ÷ 9	(13) 90 ÷ 9	(14) 54 ÷ 9	(15) 99 ÷ 9
(16) 27 ÷ 9	(17) 81 ÷ 9	(18) 9 ÷ 9	(19) 108 ÷ 9	(20) 36 ÷ 9
(21) 0 ÷ 9	(22) 18 ÷ 9	(23) 63 ÷ 9	(24) 90 ÷ 9	(25) 18 ÷ 9
(26) 63 ÷ 9	(27) 81 ÷ 9	(28) 9 ÷ 9	(29) 108 ÷ 9	(30) 0 ÷ 9
(31) 72 ÷ 9	(32) 45 ÷ 9	(33) 36 ÷ 9	(34) 27 ÷ 9	(35) 99 ÷ 9
(36) 54 ÷ 9	(37) 90 ÷ 9	(38) 81 ÷ 9	(39) 9 ÷ 9	(40) 99 ÷ 9

53

Divide by 7-9

Time: ______

Score: ______/ 40

1. 16 ÷ 8
2. 80 ÷ 8
3. 7 ÷ 7
4. 54 ÷ 9
5. 64 ÷ 8
6. 27 ÷ 9
7. 90 ÷ 9
8. 63 ÷ 7
9. 77 ÷ 7
10. 36 ÷ 9
11. 49 ÷ 7
12. 9 ÷ 9
13. 84 ÷ 7
14. 108 ÷ 9
15. 72 ÷ 9
16. 70 ÷ 7
17. 56 ÷ 8
18. 72 ÷ 8
19. 81 ÷ 9
20. 40 ÷ 8
21. 28 ÷ 7
22. 96 ÷ 8
23. 8 ÷ 8
24. 56 ÷ 7
25. 0 ÷ 8
26. 35 ÷ 7
27. 24 ÷ 8
28. 14 ÷ 7
29. 42 ÷ 7
30. 45 ÷ 9
31. 88 ÷ 8
32. 0 ÷ 9
33. 99 ÷ 9
34. 21 ÷ 7
35. 48 ÷ 8
36. 32 ÷ 8
37. 63 ÷ 9
38. 18 ÷ 9
39. 0 ÷ 7
40. 70 ÷ 7

54

Divide by 7-9

Time: __________

Score: _____/ 40

(1) 0 ÷ 8	(2) 8 ÷ 8	(3) 27 ÷ 9	(4) 81 ÷ 9	(5) 108 ÷ 9
(6) 84 ÷ 7	(7) 45 ÷ 9	(8) 14 ÷ 7	(9) 72 ÷ 8	(10) 56 ÷ 7
(11) 96 ÷ 8	(12) 80 ÷ 8	(13) 18 ÷ 9	(14) 56 ÷ 8	(15) 40 ÷ 8
(16) 49 ÷ 7	(17) 24 ÷ 8	(18) 16 ÷ 8	(19) 9 ÷ 9	(20) 35 ÷ 7
(21) 64 ÷ 8	(22) 88 ÷ 8	(23) 48 ÷ 8	(24) 63 ÷ 7	(25) 0 ÷ 9
(26) 0 ÷ 7	(27) 77 ÷ 7	(28) 28 ÷ 7	(29) 36 ÷ 9	(30) 72 ÷ 9
(31) 99 ÷ 9	(32) 7 ÷ 7	(33) 54 ÷ 9	(34) 90 ÷ 9	(35) 63 ÷ 9
(36) 21 ÷ 7	(37) 42 ÷ 7	(38) 32 ÷ 8	(39) 88 ÷ 8	(40) 28 ÷ 7

55

Divide by 7-9

Time: ________

Score: _____/ 40

(1) 8 ÷ 8	(2) 42 ÷ 7	(3) 35 ÷ 7	(4) 77 ÷ 7	(5) 7 ÷ 7
(6) 36 ÷ 9	(7) 99 ÷ 9	(8) 49 ÷ 7	(9) 40 ÷ 8	(10) 45 ÷ 9
(11) 9 ÷ 9	(12) 48 ÷ 8	(13) 21 ÷ 7	(14) 56 ÷ 8	(15) 90 ÷ 9
(16) 27 ÷ 9	(17) 0 ÷ 9	(18) 56 ÷ 7	(19) 70 ÷ 7	(20) 96 ÷ 8
(21) 72 ÷ 8	(22) 32 ÷ 8	(23) 14 ÷ 7	(24) 64 ÷ 8	(25) 108 ÷ 9
(26) 24 ÷ 8	(27) 18 ÷ 9	(28) 81 ÷ 9	(29) 84 ÷ 7	(30) 72 ÷ 9
(31) 63 ÷ 7	(32) 54 ÷ 9	(33) 0 ÷ 8	(34) 16 ÷ 8	(35) 0 ÷ 7
(36) 63 ÷ 9	(37) 80 ÷ 8	(38) 70 ÷ 7	(39) 63 ÷ 9	(40) 36 ÷ 9

56

Divide by 7-9

Time: ________

Score: ____/ 40

1. $0 \div 8$
2. $27 \div 9$
3. $0 \div 9$
4. $40 \div 8$
5. $72 \div 8$
6. $32 \div 8$
7. $45 \div 9$
8. $108 \div 9$
9. $90 \div 9$
10. $72 \div 9$
11. $7 \div 7$
12. $18 \div 9$
13. $64 \div 8$
14. $9 \div 9$
15. $24 \div 8$
16. $28 \div 7$
17. $16 \div 8$
18. $88 \div 8$
19. $81 \div 9$
20. $56 \div 7$
21. $63 \div 7$
22. $99 \div 9$
23. $21 \div 7$
24. $49 \div 7$
25. $8 \div 8$
26. $80 \div 8$
27. $35 \div 7$
28. $56 \div 8$
29. $14 \div 7$
30. $42 \div 7$
31. $77 \div 7$
32. $54 \div 9$
33. $84 \div 7$
34. $0 \div 7$
35. $96 \div 8$
36. $48 \div 8$
37. $0 \div 9$
38. $63 \div 9$
39. $24 \div 8$
40. $16 \div 8$

57

Divide by 7-9

Time: ________

Score: _____/ 40

① 45 ÷ 9	② 21 ÷ 7	③ 56 ÷ 8	④ 80 ÷ 8	⑤ 0 ÷ 8
⑥ 35 ÷ 7	⑦ 64 ÷ 8	⑧ 63 ÷ 7	⑨ 36 ÷ 9	⑩ 70 ÷ 7
⑪ 72 ÷ 8	⑫ 49 ÷ 7	⑬ 77 ÷ 7	⑭ 27 ÷ 9	⑮ 18 ÷ 9
⑯ 88 ÷ 8	⑰ 28 ÷ 7	⑱ 14 ÷ 7	⑲ 9 ÷ 9	⑳ 96 ÷ 8
㉑ 54 ÷ 9	㉒ 108 ÷ 9	㉓ 40 ÷ 8	㉔ 72 ÷ 9	㉕ 84 ÷ 7
㉖ 48 ÷ 8	㉗ 56 ÷ 7	㉘ 0 ÷ 7	㉙ 81 ÷ 9	㉚ 7 ÷ 7
㉛ 32 ÷ 8	㉜ 8 ÷ 8	㉝ 90 ÷ 9	㉞ 99 ÷ 9	㉟ 42 ÷ 7
㊱ 81 ÷ 9	㊲ 96 ÷ 8	㊳ 7 ÷ 7	㊴ 84 ÷ 7	㊵ 16 ÷ 8

58

Divide by 7-9

Time: ______

Score: _____/ 40

① 18 ÷ 9	② 49 ÷ 7	③ 28 ÷ 7	④ 54 ÷ 9	⑤ 72 ÷ 9
⑥ 14 ÷ 7	⑦ 108 ÷ 9	⑧ 90 ÷ 9	⑨ 8 ÷ 8	⑩ 63 ÷ 9
⑪ 77 ÷ 7	⑫ 70 ÷ 7	⑬ 9 ÷ 9	⑭ 0 ÷ 9	⑮ 88 ÷ 8
⑯ 27 ÷ 9	⑰ 99 ÷ 9	⑱ 72 ÷ 8	⑲ 45 ÷ 9	⑳ 63 ÷ 7
(21) 24 ÷ 8	(22) 32 ÷ 8	(23) 64 ÷ 8	(24) 56 ÷ 8	(25) 48 ÷ 8
(26) 40 ÷ 8	(27) 0 ÷ 7	(28) 21 ÷ 7	(29) 0 ÷ 8	(30) 35 ÷ 7
(31) 42 ÷ 7	(32) 80 ÷ 8	(33) 56 ÷ 7	(34) 36 ÷ 9	(35) 42 ÷ 7
(36) 0 ÷ 7	(37) 77 ÷ 7	(38) 36 ÷ 9	(39) 35 ÷ 7	(40) 0 ÷ 8

59

Divide by 7-9

Time: ________

Score: _____/ 40

(1) 54 ÷ 9	(2) 14 ÷ 7	(3) 56 ÷ 7	(4) 70 ÷ 7	(5) 90 ÷ 9
(6) 96 ÷ 8	(7) 7 ÷ 7	(8) 81 ÷ 9	(9) 84 ÷ 7	(10) 27 ÷ 9
(11) 80 ÷ 8	(12) 63 ÷ 9	(13) 72 ÷ 8	(14) 99 ÷ 9	(15) 56 ÷ 8
(16) 9 ÷ 9	(17) 16 ÷ 8	(18) 18 ÷ 9	(19) 40 ÷ 8	(20) 0 ÷ 9
(21) 24 ÷ 8	(22) 48 ÷ 8	(23) 72 ÷ 9	(24) 45 ÷ 9	(25) 88 ÷ 8
(26) 8 ÷ 8	(27) 32 ÷ 8	(28) 64 ÷ 8	(29) 28 ÷ 7	(30) 108 ÷ 9
(31) 21 ÷ 7	(32) 63 ÷ 7	(33) 49 ÷ 7	(34) 18 ÷ 9	(35) 64 ÷ 8
(36) 14 ÷ 7	(37) 72 ÷ 9	(38) 35 ÷ 7	(39) 90 ÷ 9	(40) 54 ÷ 9

60

Divide by 7-9

Time: ___________

Score: _____/ 40

1. $80 \div 8$
2. $16 \div 8$
3. $21 \div 7$
4. $0 \div 8$
5. $7 \div 7$
6. $63 \div 9$
7. $8 \div 8$
8. $108 \div 9$
9. $63 \div 7$
10. $0 \div 9$
11. $27 \div 9$
12. $24 \div 8$
13. $96 \div 8$
14. $70 \div 7$
15. $45 \div 9$
16. $42 \div 7$
17. $99 \div 9$
18. $9 \div 9$
19. $77 \div 7$
20. $28 \div 7$
21. $32 \div 8$
22. $49 \div 7$
23. $88 \div 8$
24. $72 \div 8$
25. $40 \div 8$
26. $84 \div 7$
27. $56 \div 7$
28. $0 \div 7$
29. $48 \div 8$
30. $56 \div 8$
31. $81 \div 9$
32. $36 \div 9$
33. $7 \div 7$
34. $0 \div 9$
35. $24 \div 8$
36. $8 \div 8$
37. $80 \div 8$
38. $18 \div 9$
39. $54 \div 9$
40. $56 \div 7$

61

Divide by 10

Time: ________

Score: _____/ 40

(1) $80 \div 10$	(2) $40 \div 10$	(3) $0 \div 10$	(4) $90 \div 10$	(5) $120 \div 10$
(6) $100 \div 10$	(7) $50 \div 10$	(8) $10 \div 10$	(9) $20 \div 10$	(10) $110 \div 10$
(11) $70 \div 10$	(12) $30 \div 10$	(13) $60 \div 10$	(14) $10 \div 10$	(15) $80 \div 10$
(16) $70 \div 10$	(17) $90 \div 10$	(18) $60 \div 10$	(19) $0 \div 10$	(20) $30 \div 10$
(21) $20 \div 10$	(22) $110 \div 10$	(23) $100 \div 10$	(24) $120 \div 10$	(25) $40 \div 10$
(26) $50 \div 10$	(27) $120 \div 10$	(28) $70 \div 10$	(29) $60 \div 10$	(30) $90 \div 10$
(31) $0 \div 10$	(32) $30 \div 10$	(33) $80 \div 10$	(34) $20 \div 10$	(35) $100 \div 10$
(36) $50 \div 10$	(37) $10 \div 10$	(38) $40 \div 10$	(39) $110 \div 10$	(40) $120 \div 10$

62

Divide by 10

Time: ___________

Score: ______/ 40

(1) 100 ÷ 10	(2) 60 ÷ 10	(3) 50 ÷ 10	(4) 0 ÷ 10	(5) 80 ÷ 10
(6) 70 ÷ 10	(7) 90 ÷ 10	(8) 110 ÷ 10	(9) 20 ÷ 10	(10) 10 ÷ 10
(11) 30 ÷ 10	(12) 40 ÷ 10	(13) 50 ÷ 10	(14) 60 ÷ 10	(15) 10 ÷ 10
(16) 70 ÷ 10	(17) 0 ÷ 10	(18) 40 ÷ 10	(19) 120 ÷ 10	(20) 100 ÷ 10
(21) 80 ÷ 10	(22) 20 ÷ 10	(23) 110 ÷ 10	(24) 30 ÷ 10	(25) 90 ÷ 10
(26) 100 ÷ 10	(27) 40 ÷ 10	(28) 70 ÷ 10	(29) 80 ÷ 10	(30) 50 ÷ 10
(31) 10 ÷ 10	(32) 30 ÷ 10	(33) 0 ÷ 10	(34) 110 ÷ 10	(35) 120 ÷ 10
(36) 20 ÷ 10	(37) 60 ÷ 10	(38) 90 ÷ 10	(39) 80 ÷ 10	(40) 20 ÷ 10

63

Divide by 10

Time: ____________

Score: _____/ 40

(1) 10 ÷ 10	(2) 30 ÷ 10	(3) 0 ÷ 10	(4) 120 ÷ 10	(5) 100 ÷ 10
(6) 90 ÷ 10	(7) 60 ÷ 10	(8) 50 ÷ 10	(9) 40 ÷ 10	(10) 70 ÷ 10
(11) 110 ÷ 10	(12) 0 ÷ 10	(13) 40 ÷ 10	(14) 50 ÷ 10	(15) 80 ÷ 10
(16) 20 ÷ 10	(17) 110 ÷ 10	(18) 100 ÷ 10	(19) 120 ÷ 10	(20) 70 ÷ 10
(21) 60 ÷ 10	(22) 30 ÷ 10	(23) 10 ÷ 10	(24) 90 ÷ 10	(25) 80 ÷ 10
(26) 90 ÷ 10	(27) 60 ÷ 10	(28) 40 ÷ 10	(29) 10 ÷ 10	(30) 120 ÷ 10
(31) 50 ÷ 10	(32) 100 ÷ 10	(33) 30 ÷ 10	(34) 70 ÷ 10	(35) 20 ÷ 10
(36) 110 ÷ 10	(37) 0 ÷ 10	(38) 100 ÷ 10	(39) 70 ÷ 10	(40) 110 ÷ 10

64

Divide by 10

Time: ___________

Score: _____/ 40

(1) $20 \div 10$	(2) $50 \div 10$	(3) $90 \div 10$	(4) $30 \div 10$	(5) $120 \div 10$
(6) $80 \div 10$	(7) $0 \div 10$	(8) $60 \div 10$	(9) $40 \div 10$	(10) $10 \div 10$
(11) $20 \div 10$	(12) $70 \div 10$	(13) $10 \div 10$	(14) $80 \div 10$	(15) $60 \div 10$
(16) $0 \div 10$	(17) $110 \div 10$	(18) $40 \div 10$	(19) $100 \div 10$	(20) $120 \div 10$
(21) $50 \div 10$	(22) $90 \div 10$	(23) $30 \div 10$	(24) $70 \div 10$	(25) $30 \div 10$
(26) $100 \div 10$	(27) $110 \div 10$	(28) $20 \div 10$	(29) $50 \div 10$	(30) $10 \div 10$
(31) $80 \div 10$	(32) $120 \div 10$	(33) $60 \div 10$	(34) $90 \div 10$	(35) $0 \div 10$
(36) $40 \div 10$	(37) $90 \div 10$	(38) $30 \div 10$	(39) $110 \div 10$	(40) $20 \div 10$

65

Divide by 11

Time: ________

Score: ____/ 40

① 55 ÷ 11	② 99 ÷ 11	③ 11 ÷ 11	④ 33 ÷ 11	⑤ 77 ÷ 11
⑥ 66 ÷ 11	⑦ 44 ÷ 11	⑧ 110 ÷ 11	⑨ 88 ÷ 11	⑩ 132 ÷ 11
⑪ 121 ÷ 11	⑫ 0 ÷ 11	⑬ 22 ÷ 11	⑭ 132 ÷ 11	⑮ 121 ÷ 11
⑯ 44 ÷ 11	⑰ 55 ÷ 11	⑱ 110 ÷ 11	⑲ 99 ÷ 11	⑳ 11 ÷ 11
(21) 88 ÷ 11	(22) 33 ÷ 11	(23) 66 ÷ 11	(24) 77 ÷ 11	(25) 22 ÷ 11
(26) 0 ÷ 11	(27) 33 ÷ 11	(28) 44 ÷ 11	(29) 66 ÷ 11	(30) 22 ÷ 11
(31) 0 ÷ 11	(32) 11 ÷ 11	(33) 132 ÷ 11	(34) 55 ÷ 11	(35) 88 ÷ 11
(36) 99 ÷ 11	(37) 110 ÷ 11	(38) 77 ÷ 11	(39) 121 ÷ 11	(40) 44 ÷ 11

66

Divide by 11

Time: ________

Score: _____/ 40

(1) $121 \div 11$	(2) $33 \div 11$	(3) $88 \div 11$	(4) $110 \div 11$	(5) $11 \div 11$
(6) $99 \div 11$	(7) $77 \div 11$	(8) $0 \div 11$	(9) $55 \div 11$	(10) $66 \div 11$
(11) $132 \div 11$	(12) $22 \div 11$	(13) $55 \div 11$	(14) $132 \div 11$	(15) $66 \div 11$
(16) $110 \div 11$	(17) $88 \div 11$	(18) $77 \div 11$	(19) $11 \div 11$	(20) $121 \div 11$
(21) $0 \div 11$	(22) $99 \div 11$	(23) $22 \div 11$	(24) $44 \div 11$	(25) $33 \div 11$
(26) $132 \div 11$	(27) $110 \div 11$	(28) $44 \div 11$	(29) $55 \div 11$	(30) $33 \div 11$
(31) $11 \div 11$	(32) $22 \div 11$	(33) $77 \div 11$	(34) $0 \div 11$	(35) $66 \div 11$
(36) $121 \div 11$	(37) $88 \div 11$	(38) $99 \div 11$	(39) $55 \div 11$	(40) $132 \div 11$

67

Divide by 11

Time: ________

Score: _____/ 40

(1) 99 ÷ 11	(2) 66 ÷ 11	(3) 121 ÷ 11	(4) 44 ÷ 11	(5) 11 ÷ 11
(6) 0 ÷ 11	(7) 77 ÷ 11	(8) 22 ÷ 11	(9) 88 ÷ 11	(10) 110 ÷ 11
(11) 33 ÷ 11	(12) 66 ÷ 11	(13) 22 ÷ 11	(14) 0 ÷ 11	(15) 132 ÷ 11
(16) 99 ÷ 11	(17) 55 ÷ 11	(18) 44 ÷ 11	(19) 11 ÷ 11	(20) 33 ÷ 11
(21) 121 ÷ 11	(22) 88 ÷ 11	(23) 77 ÷ 11	(24) 110 ÷ 11	(25) 77 ÷ 11
(26) 66 ÷ 11	(27) 99 ÷ 11	(28) 88 ÷ 11	(29) 132 ÷ 11	(30) 22 ÷ 11
(31) 33 ÷ 11	(32) 121 ÷ 11	(33) 55 ÷ 11	(34) 110 ÷ 11	(35) 11 ÷ 11
(36) 0 ÷ 11	(37) 44 ÷ 11	(38) 121 ÷ 11	(39) 99 ÷ 11	(40) 0 ÷ 11

68

Divide by 11

Time: ________

Score: _____/ 40

(1) $33 \div 11$	(2) $22 \div 11$	(3) $11 \div 11$	(4) $44 \div 11$	(5) $88 \div 11$
(6) $77 \div 11$	(7) $66 \div 11$	(8) $55 \div 11$	(9) $132 \div 11$	(10) $110 \div 11$
(11) $66 \div 11$	(12) $0 \div 11$	(13) $110 \div 11$	(14) $11 \div 11$	(15) $132 \div 11$
(16) $88 \div 11$	(17) $44 \div 11$	(18) $99 \div 11$	(19) $55 \div 11$	(20) $77 \div 11$
(21) $33 \div 11$	(22) $121 \div 11$	(23) $22 \div 11$	(24) $55 \div 11$	(25) $88 \div 11$
(26) $11 \div 11$	(27) $77 \div 11$	(28) $66 \div 11$	(29) $0 \div 11$	(30) $121 \div 11$
(31) $132 \div 11$	(32) $33 \div 11$	(33) $22 \div 11$	(34) $110 \div 11$	(35) $44 \div 11$
(36) $99 \div 11$	(37) $121 \div 11$	(38) $55 \div 11$	(39) $44 \div 11$	(40) $132 \div 11$

69

Divide by 12

Time: ________

Score: _____/ 40

① 96 ÷ 12 = ____	② 84 ÷ 12 = ____	③ 24 ÷ 12 = ____	④ 0 ÷ 12 = ____	⑤ 120 ÷ 12 = ____
⑥ 144 ÷ 12 = ____	⑦ 48 ÷ 12 = ____	⑧ 60 ÷ 12 = ____	⑨ 108 ÷ 12 = ____	⑩ 36 ÷ 12 = ____
⑪ 12 ÷ 12 = ____	⑫ 132 ÷ 12 = ____	⑬ 72 ÷ 12 = ____	⑭ 0 ÷ 12 = ____	⑮ 84 ÷ 12 = ____
⑯ 144 ÷ 12 = ____	⑰ 36 ÷ 12 = ____	⑱ 72 ÷ 12 = ____	⑲ 120 ÷ 12 = ____	⑳ 108 ÷ 12 = ____
(21) 48 ÷ 12 = ____	(22) 12 ÷ 12 = ____	(23) 24 ÷ 12 = ____	(24) 60 ÷ 12 = ____	(25) 132 ÷ 12 = ____
(26) 96 ÷ 12 = ____	(27) 36 ÷ 12 = ____	(28) 132 ÷ 12 = ____	(29) 84 ÷ 12 = ____	(30) 48 ÷ 12 = ____
(31) 0 ÷ 12 = ____	(32) 120 ÷ 12 = ____	(33) 144 ÷ 12 = ____	(34) 96 ÷ 12 = ____	(35) 72 ÷ 12 = ____
(36) 12 ÷ 12 = ____	(37) 24 ÷ 12 = ____	(38) 108 ÷ 12 = ____	(39) 60 ÷ 12 = ____	(40) 108 ÷ 12 = ____

70

Divide by 12

Time: ________

Score: _____/ 40

(1) 132 ÷ 12	(2) 48 ÷ 12	(3) 72 ÷ 12	(4) 120 ÷ 12	(5) 24 ÷ 12
(6) 144 ÷ 12	(7) 96 ÷ 12	(8) 12 ÷ 12	(9) 0 ÷ 12	(10) 84 ÷ 12
(11) 36 ÷ 12	(12) 60 ÷ 12	(13) 72 ÷ 12	(14) 60 ÷ 12	(15) 36 ÷ 12
(16) 84 ÷ 12	(17) 144 ÷ 12	(18) 120 ÷ 12	(19) 132 ÷ 12	(20) 24 ÷ 12
(21) 0 ÷ 12	(22) 48 ÷ 12	(23) 12 ÷ 12	(24) 108 ÷ 12	(25) 96 ÷ 12
(26) 12 ÷ 12	(27) 48 ÷ 12	(28) 96 ÷ 12	(29) 24 ÷ 12	(30) 108 ÷ 12
(31) 72 ÷ 12	(32) 120 ÷ 12	(33) 132 ÷ 12	(34) 144 ÷ 12	(35) 0 ÷ 12
(36) 84 ÷ 12	(37) 36 ÷ 12	(38) 60 ÷ 12	(39) 120 ÷ 12	(40) 36 ÷ 12

71

Divide by 12

Time: __________

Score: _____/ 40

(1) $60 \div 12$	(2) $132 \div 12$	(3) $108 \div 12$	(4) $72 \div 12$	(5) $24 \div 12$
(6) $48 \div 12$	(7) $12 \div 12$	(8) $84 \div 12$	(9) $96 \div 12$	(10) $144 \div 12$
(11) $0 \div 12$	(12) $144 \div 12$	(13) $24 \div 12$	(14) $132 \div 12$	(15) $108 \div 12$
(16) $120 \div 12$	(17) $12 \div 12$	(18) $72 \div 12$	(19) $36 \div 12$	(20) $96 \div 12$
(21) $48 \div 12$	(22) $60 \div 12$	(23) $0 \div 12$	(24) $84 \div 12$	(25) $72 \div 12$
(26) $0 \div 12$	(27) $24 \div 12$	(28) $12 \div 12$	(29) $132 \div 12$	(30) $144 \div 12$
(31) $120 \div 12$	(32) $108 \div 12$	(33) $60 \div 12$	(34) $36 \div 12$	(35) $96 \div 12$
(36) $48 \div 12$	(37) $84 \div 12$	(38) $96 \div 12$	(39) $0 \div 12$	(40) $72 \div 12$

72

Divide by 12

Time: ________

Score: _____/ 40

(1) 48 ÷ 12	(2) 120 ÷ 12	(3) 144 ÷ 12	(4) 60 ÷ 12	(5) 24 ÷ 12
(6) 36 ÷ 12	(7) 108 ÷ 12	(8) 132 ÷ 12	(9) 84 ÷ 12	(10) 12 ÷ 12
(11) 108 ÷ 12	(12) 60 ÷ 12	(13) 0 ÷ 12	(14) 120 ÷ 12	(15) 72 ÷ 12
(16) 36 ÷ 12	(17) 84 ÷ 12	(18) 24 ÷ 12	(19) 96 ÷ 12	(20) 12 ÷ 12
(21) 48 ÷ 12	(22) 132 ÷ 12	(23) 144 ÷ 12	(24) 0 ÷ 12	(25) 12 ÷ 12
(26) 60 ÷ 12	(27) 84 ÷ 12	(28) 132 ÷ 12	(29) 48 ÷ 12	(30) 96 ÷ 12
(31) 144 ÷ 12	(32) 24 ÷ 12	(33) 108 ÷ 12	(34) 120 ÷ 12	(35) 36 ÷ 12
(36) 72 ÷ 12	(37) 132 ÷ 12	(38) 36 ÷ 12	(39) 72 ÷ 12	(40) 96 ÷ 12

73

Divide by 10-12

Time: __________

Score: _____/ 40

1. 22 ÷ 11
2. 120 ÷ 12
3. 50 ÷ 10
4. 88 ÷ 11
5. 72 ÷ 12
6. 10 ÷ 10
7. 12 ÷ 12
8. 60 ÷ 12
9. 108 ÷ 12
10. 33 ÷ 11
11. 0 ÷ 11
12. 66 ÷ 11
13. 99 ÷ 11
14. 120 ÷ 10
15. 55 ÷ 11
16. 96 ÷ 12
17. 84 ÷ 12
18. 11 ÷ 11
19. 44 ÷ 11
20. 110 ÷ 11
21. 90 ÷ 10
22. 144 ÷ 12
23. 70 ÷ 10
24. 0 ÷ 12
25. 24 ÷ 12
26. 121 ÷ 11
27. 110 ÷ 10
28. 100 ÷ 10
29. 80 ÷ 10
30. 60 ÷ 10
31. 30 ÷ 10
32. 48 ÷ 12
33. 132 ÷ 11
34. 132 ÷ 12
35. 20 ÷ 10
36. 0 ÷ 10
37. 77 ÷ 11
38. 36 ÷ 12
39. 40 ÷ 10
40. 60 ÷ 10

74

Divide by 10-12

Time: ____________

Score: ______/ 40

1. 120 ÷ 10
2. 100 ÷ 10
3. 132 ÷ 12
4. 110 ÷ 11
5. 50 ÷ 10
6. 144 ÷ 12
7. 77 ÷ 11
8. 132 ÷ 11
9. 72 ÷ 12
10. 0 ÷ 11
11. 99 ÷ 11
12. 60 ÷ 12
13. 24 ÷ 12
14. 22 ÷ 11
15. 0 ÷ 10
16. 120 ÷ 12
17. 121 ÷ 11
18. 0 ÷ 12
19. 108 ÷ 12
20. 33 ÷ 11
21. 40 ÷ 10
22. 44 ÷ 11
23. 55 ÷ 11
24. 110 ÷ 10
25. 10 ÷ 10
26. 36 ÷ 12
27. 80 ÷ 10
28. 88 ÷ 11
29. 70 ÷ 10
30. 90 ÷ 10
31. 66 ÷ 11
32. 11 ÷ 11
33. 12 ÷ 12
34. 84 ÷ 12
35. 48 ÷ 12
36. 30 ÷ 10
37. 96 ÷ 12
38. 20 ÷ 10
39. 0 ÷ 12
40. 22 ÷ 11

75

Divide by 10-12

Time: __________

Score: _____/ 40

(1) 99 ÷ 11	(2) 132 ÷ 11	(3) 108 ÷ 12	(4) 120 ÷ 10	(5) 100 ÷ 10
(6) 33 ÷ 11	(7) 50 ÷ 10	(8) 55 ÷ 11	(9) 77 ÷ 11	(10) 90 ÷ 10
(11) 0 ÷ 10	(12) 10 ÷ 10	(13) 72 ÷ 12	(14) 0 ÷ 11	(15) 70 ÷ 10
(16) 96 ÷ 12	(17) 48 ÷ 12	(18) 121 ÷ 11	(19) 110 ÷ 11	(20) 20 ÷ 10
(21) 88 ÷ 11	(22) 110 ÷ 10	(23) 30 ÷ 10	(24) 60 ÷ 10	(25) 40 ÷ 10
(26) 60 ÷ 12	(27) 144 ÷ 12	(28) 24 ÷ 12	(29) 44 ÷ 11	(30) 84 ÷ 12
(31) 66 ÷ 11	(32) 11 ÷ 11	(33) 132 ÷ 12	(34) 12 ÷ 12	(35) 120 ÷ 12
(36) 36 ÷ 12	(37) 80 ÷ 10	(38) 44 ÷ 11	(39) 48 ÷ 12	(40) 96 ÷ 12

76

Divide by 10-12

Time: ________

Score: _____/ 40

(1) 24 ÷ 12	(2) 72 ÷ 12	(3) 100 ÷ 10	(4) 20 ÷ 10	(5) 11 ÷ 11
(6) 36 ÷ 12	(7) 144 ÷ 12	(8) 66 ÷ 11	(9) 60 ÷ 10	(10) 99 ÷ 11
(11) 70 ÷ 10	(12) 90 ÷ 10	(13) 50 ÷ 10	(14) 132 ÷ 11	(15) 110 ÷ 11
(16) 120 ÷ 10	(17) 88 ÷ 11	(18) 22 ÷ 11	(19) 10 ÷ 10	(20) 84 ÷ 12
(21) 30 ÷ 10	(22) 0 ÷ 10	(23) 40 ÷ 10	(24) 0 ÷ 11	(25) 33 ÷ 11
(26) 80 ÷ 10	(27) 108 ÷ 12	(28) 55 ÷ 11	(29) 12 ÷ 12	(30) 120 ÷ 12
(31) 77 ÷ 11	(32) 121 ÷ 11	(33) 60 ÷ 12	(34) 0 ÷ 12	(35) 132 ÷ 12
(36) 110 ÷ 10	(37) 44 ÷ 11	(38) 100 ÷ 10	(39) 99 ÷ 11	(40) 121 ÷ 11

77

Divide by 10-12

Time: ________

Score: _____/ 40

(1) 132 ÷ 11	(2) 40 ÷ 10	(3) 30 ÷ 10	(4) 36 ÷ 12	(5) 11 ÷ 11
(6) 33 ÷ 11	(7) 55 ÷ 11	(8) 22 ÷ 11	(9) 120 ÷ 10	(10) 60 ÷ 10
(11) 144 ÷ 12	(12) 24 ÷ 12	(13) 110 ÷ 11	(14) 90 ÷ 10	(15) 110 ÷ 10
(16) 108 ÷ 12	(17) 66 ÷ 11	(18) 0 ÷ 12	(19) 84 ÷ 12	(20) 77 ÷ 11
(21) 20 ÷ 10	(22) 0 ÷ 11	(23) 10 ÷ 10	(24) 72 ÷ 12	(25) 48 ÷ 12
(26) 12 ÷ 12	(27) 60 ÷ 12	(28) 88 ÷ 11	(29) 70 ÷ 10	(30) 132 ÷ 12
(31) 120 ÷ 12	(32) 96 ÷ 12	(33) 50 ÷ 10	(34) 80 ÷ 10	(35) 0 ÷ 10
(36) 0 ÷ 11	(37) 50 ÷ 10	(38) 22 ÷ 11	(39) 10 ÷ 10	(40) 70 ÷ 10

78

Divide by 10-12

Time: __________

Score: _____/ 40

(1) $24 \div 12$	(2) $12 \div 12$	(3) $120 \div 10$	(4) $36 \div 12$	(5) $121 \div 11$
(6) $30 \div 10$	(7) $132 \div 12$	(8) $100 \div 10$	(9) $48 \div 12$	(10) $0 \div 12$
(11) $144 \div 12$	(12) $96 \div 12$	(13) $99 \div 11$	(14) $66 \div 11$	(15) $120 \div 12$
(16) $88 \div 11$	(17) $84 \div 12$	(18) $33 \div 11$	(19) $40 \div 10$	(20) $20 \div 10$
(21) $55 \div 11$	(22) $44 \div 11$	(23) $60 \div 12$	(24) $11 \div 11$	(25) $110 \div 10$
(26) $80 \div 10$	(27) $77 \div 11$	(28) $90 \div 10$	(29) $110 \div 11$	(30) $72 \div 12$
(31) $60 \div 10$	(32) $0 \div 10$	(33) $108 \div 12$	(34) $132 \div 11$	(35) $0 \div 10$
(36) $22 \div 11$	(37) $110 \div 11$	(38) $12 \div 12$	(39) $11 \div 11$	(40) $132 \div 11$

79

Divide by 10-12

Time: ________

Score: _____/ 40

(1) 96 ÷ 12	(2) 20 ÷ 10	(3) 70 ÷ 10	(4) 0 ÷ 11	(5) 121 ÷ 11
(6) 77 ÷ 11	(7) 120 ÷ 10	(8) 0 ÷ 12	(9) 36 ÷ 12	(10) 100 ÷ 10
(11) 60 ÷ 12	(12) 40 ÷ 10	(13) 10 ÷ 10	(14) 132 ÷ 12	(15) 90 ÷ 10
(16) 50 ÷ 10	(17) 88 ÷ 11	(18) 33 ÷ 11	(19) 72 ÷ 12	(20) 99 ÷ 11
(21) 44 ÷ 11	(22) 110 ÷ 10	(23) 30 ÷ 10	(24) 24 ÷ 12	(25) 144 ÷ 12
(26) 120 ÷ 12	(27) 108 ÷ 12	(28) 48 ÷ 12	(29) 80 ÷ 10	(30) 60 ÷ 10
(31) 55 ÷ 11	(32) 84 ÷ 12	(33) 66 ÷ 11	(34) 108 ÷ 12	(35) 11 ÷ 11
(36) 12 ÷ 12	(37) 72 ÷ 12	(38) 110 ÷ 10	(39) 66 ÷ 11	(40) 44 ÷ 11

80

Divide by 10-12

Time: ________

Score: _____/ 40

(1) 77 ÷ 11	(2) 48 ÷ 12	(3) 55 ÷ 11	(4) 60 ÷ 12	(5) 110 ÷ 11
(6) 30 ÷ 10	(7) 99 ÷ 11	(8) 84 ÷ 12	(9) 100 ÷ 10	(10) 0 ÷ 12
(11) 33 ÷ 11	(12) 120 ÷ 12	(13) 10 ÷ 10	(14) 80 ÷ 10	(15) 121 ÷ 11
(16) 20 ÷ 10	(17) 90 ÷ 10	(18) 60 ÷ 10	(19) 24 ÷ 12	(20) 144 ÷ 12
(21) 120 ÷ 10	(22) 0 ÷ 11	(23) 0 ÷ 10	(24) 88 ÷ 11	(25) 50 ÷ 10
(26) 96 ÷ 12	(27) 40 ÷ 10	(28) 132 ÷ 12	(29) 70 ÷ 10	(30) 36 ÷ 12
(31) 22 ÷ 11	(32) 132 ÷ 11	(33) 144 ÷ 12	(34) 100 ÷ 10	(35) 120 ÷ 10
(36) 110 ÷ 11	(37) 132 ÷ 11	(38) 11 ÷ 11	(39) 33 ÷ 11	(40) 99 ÷ 11

81

Divide by 1-12

Time: ________

Score: _____/ 40

1. $30 \div 5$
2. $121 \div 11$
3. $5 \div 1$
4. $120 \div 12$
5. $7 \div 7$
6. $50 \div 5$
7. $36 \div 12$
8. $42 \div 7$
9. $4 \div 4$
10. $10 \div 5$
11. $35 \div 5$
12. $2 \div 1$
13. $30 \div 3$
14. $48 \div 6$
15. $24 \div 4$
16. $55 \div 11$
17. $22 \div 11$
18. $11 \div 1$
19. $96 \div 12$
20. $12 \div 1$
21. $44 \div 11$
22. $60 \div 12$
23. $15 \div 3$
24. $40 \div 10$
25. $6 \div 3$
26. $9 \div 9$
27. $8 \div 4$
28. $70 \div 10$
29. $7 \div 1$
30. $35 \div 7$
31. $30 \div 6$
32. $25 \div 5$
33. $48 \div 8$
34. $24 \div 2$
35. $60 \div 6$
36. $66 \div 6$
37. $40 \div 5$
38. $45 \div 9$
39. $96 \div 8$
40. $60 \div 5$

82

Divide by 1-12

Time: ___________

Score: _____/ 40

1. 0 ÷ 2
2. 77 ÷ 11
3. 110 ÷ 11
4. 3 ÷ 1
5. 2 ÷ 2
6. 0 ÷ 5
7. 0 ÷ 10
8. 18 ÷ 3
9. 21 ÷ 3
10. 32 ÷ 8
11. 0 ÷ 1
12. 27 ÷ 3
13. 14 ÷ 2
14. 0 ÷ 8
15. 22 ÷ 2
16. 27 ÷ 9
17. 77 ÷ 7
18. 16 ÷ 4
19. 81 ÷ 9
20. 120 ÷ 10
21. 72 ÷ 8
22. 20 ÷ 2
23. 28 ÷ 7
24. 0 ÷ 7
25. 6 ÷ 1
26. 14 ÷ 7
27. 72 ÷ 6
28. 10 ÷ 1
29. 0 ÷ 3
30. 10 ÷ 10
31. 16 ÷ 2
32. 72 ÷ 12
33. 36 ÷ 6
34. 28 ÷ 4
35. 36 ÷ 3
36. 20 ÷ 10
37. 15 ÷ 5
38. 30 ÷ 10
39. 12 ÷ 2
40. 18 ÷ 6

83

Divide by 1-12

Time: __________

Score: _____/ 40

(1) 32 ÷ 4	(2) 0 ÷ 4	(3) 49 ÷ 7	(4) 40 ÷ 8	(5) 6 ÷ 6
(6) 56 ÷ 7	(7) 12 ÷ 4	(8) 63 ÷ 9	(9) 0 ÷ 11	(10) 12 ÷ 12
(11) 24 ÷ 12	(12) 48 ÷ 4	(13) 99 ÷ 9	(14) 48 ÷ 12	(15) 66 ÷ 11
(16) 24 ÷ 3	(17) 36 ÷ 4	(18) 80 ÷ 8	(19) 24 ÷ 6	(20) 9 ÷ 3
(21) 132 ÷ 12	(22) 24 ÷ 8	(23) 6 ÷ 2	(24) 5 ÷ 5	(25) 0 ÷ 6
(26) 3 ÷ 3	(27) 18 ÷ 2	(28) 84 ÷ 12	(29) 132 ÷ 11	(30) 9 ÷ 1
(31) 88 ÷ 11	(32) 4 ÷ 1	(33) 8 ÷ 8	(34) 50 ÷ 10	(35) 70 ÷ 7
(36) 12 ÷ 3	(37) 84 ÷ 7	(38) 12 ÷ 6	(39) 8 ÷ 1	(40) 100 ÷ 10

84

Divide by 1-12

Time: __________

Score: _____/ 40

(1) 11 ÷ 11	(2) 54 ÷ 9	(3) 4 ÷ 2	(4) 20 ÷ 5	(5) 33 ÷ 11
(6) 110 ÷ 10	(7) 1 ÷ 1	(8) 60 ÷ 10	(9) 90 ÷ 10	(10) 0 ÷ 9
(11) 99 ÷ 11	(12) 56 ÷ 8	(13) 64 ÷ 8	(14) 63 ÷ 7	(15) 44 ÷ 4
(16) 20 ÷ 4	(17) 80 ÷ 10	(18) 40 ÷ 4	(19) 0 ÷ 12	(20) 108 ÷ 9
(21) 90 ÷ 9	(22) 33 ÷ 3	(23) 21 ÷ 7	(24) 72 ÷ 9	(25) 10 ÷ 2
(26) 108 ÷ 12	(27) 55 ÷ 5	(28) 16 ÷ 8	(29) 36 ÷ 9	(30) 54 ÷ 6
(31) 18 ÷ 9	(32) 144 ÷ 12	(33) 88 ÷ 8	(34) 8 ÷ 2	(35) 45 ÷ 5
(36) 42 ÷ 6	(37) 84 ÷ 12	(38) 21 ÷ 7	(39) 0 ÷ 11	(40) 28 ÷ 4

85

Divide by 1-12

Time: ________

Score: _____/ 40

(1) 121 ÷ 11	(2) 48 ÷ 6	(3) 4 ÷ 4	(4) 144 ÷ 12	(5) 60 ÷ 12
(6) 0 ÷ 6	(7) 110 ÷ 10	(8) 55 ÷ 5	(9) 10 ÷ 5	(10) 66 ÷ 6
(11) 14 ÷ 2	(12) 60 ÷ 10	(13) 11 ÷ 11	(14) 3 ÷ 1	(15) 72 ÷ 8
(16) 0 ÷ 3	(17) 2 ÷ 1	(18) 54 ÷ 9	(19) 27 ÷ 3	(20) 72 ÷ 6
(21) 9 ÷ 3	(22) 63 ÷ 7	(23) 48 ÷ 12	(24) 36 ÷ 9	(25) 5 ÷ 1
(26) 0 ÷ 5	(27) 9 ÷ 1	(28) 24 ÷ 2	(29) 90 ÷ 10	(30) 18 ÷ 3
(31) 32 ÷ 8	(32) 40 ÷ 5	(33) 42 ÷ 7	(34) 2 ÷ 2	(35) 4 ÷ 1
(36) 0 ÷ 4	(37) 60 ÷ 6	(38) 33 ÷ 3	(39) 77 ÷ 7	(40) 12 ÷ 12

86

Divide by 1-12

Time: ________

Score: _____/ 40

(1) $12 \div 3$	(2) $10 \div 2$	(3) $5 \div 5$	(4) $8 \div 4$	(5) $48 \div 4$
(6) $81 \div 9$	(7) $90 \div 9$	(8) $21 \div 3$	(9) $70 \div 7$	(10) $30 \div 10$
(11) $40 \div 10$	(12) $0 \div 8$	(13) $132 \div 12$	(14) $0 \div 9$	(15) $15 \div 5$
(16) $12 \div 1$	(17) $35 \div 5$	(18) $24 \div 8$	(19) $120 \div 12$	(20) $16 \div 2$
(21) $96 \div 12$	(22) $30 \div 5$	(23) $40 \div 8$	(24) $24 \div 6$	(25) $15 \div 3$
(26) $12 \div 6$	(27) $0 \div 12$	(28) $132 \div 11$	(29) $70 \div 10$	(30) $110 \div 11$
(31) $45 \div 9$	(32) $63 \div 9$	(33) $0 \div 2$	(34) $20 \div 2$	(35) $55 \div 11$
(36) $35 \div 7$	(37) $108 \div 9$	(38) $8 \div 8$	(39) $8 \div 1$	(40) $49 \div 7$

87

Divide by 1-12

Time: ________

Score: _____/ 40

(1) $24 \div 3$	(2) $36 \div 4$	(3) $10 \div 1$	(4) $14 \div 7$	(5) $30 \div 3$
(6) $80 \div 8$	(7) $12 \div 4$	(8) $100 \div 10$	(9) $6 \div 2$	(10) $28 \div 7$
(11) $36 \div 3$	(12) $48 \div 8$	(13) $30 \div 6$	(14) $64 \div 8$	(15) $44 \div 4$
(16) $96 \div 8$	(17) $18 \div 6$	(18) $36 \div 6$	(19) $0 \div 1$	(20) $27 \div 9$
(21) $44 \div 11$	(22) $16 \div 8$	(23) $20 \div 4$	(24) $50 \div 10$	(25) $42 \div 6$
(26) $24 \div 12$	(27) $80 \div 10$	(28) $18 \div 2$	(29) $0 \div 10$	(30) $33 \div 11$
(31) $4 \div 2$	(32) $66 \div 11$	(33) $22 \div 11$	(34) $6 \div 6$	(35) $88 \div 8$
(36) $60 \div 5$	(37) $40 \div 4$	(38) $25 \div 5$	(39) $32 \div 4$	(40) $6 \div 1$

88

Divide by 1-12

Time: __________

Score: _____/ 40

(1) $6 \div 3$	(2) $20 \div 10$	(3) $22 \div 2$	(4) $45 \div 5$	(5) $72 \div 12$
(6) $50 \div 5$	(7) $9 \div 9$	(8) $108 \div 12$	(9) $99 \div 9$	(10) $20 \div 5$
(11) $16 \div 4$	(12) $12 \div 2$	(13) $120 \div 10$	(14) $88 \div 11$	(15) $77 \div 11$
(16) $24 \div 4$	(17) $18 \div 9$	(18) $84 \div 7$	(19) $54 \div 6$	(20) $3 \div 3$
(21) $8 \div 2$	(22) $7 \div 7$	(23) $1 \div 1$	(24) $72 \div 9$	(25) $11 \div 1$
(26) $99 \div 11$	(27) $56 \div 8$	(28) $36 \div 12$	(29) $7 \div 1$	(30) $0 \div 7$
(31) $10 \div 10$	(32) $56 \div 7$	(33) $10 \div 1$	(34) $6 \div 2$	(35) $20 \div 10$
(36) $0 \div 8$	(37) $11 \div 11$	(38) $60 \div 10$	(39) $72 \div 8$	(40) $36 \div 3$

89

Divide by 1-12

Time: ________

Score: _____/ 40

(1) $33 \div 3$	(2) $18 \div 6$	(3) $35 \div 5$	(4) $66 \div 11$	(5) $56 \div 8$
(6) $6 \div 1$	(7) $30 \div 5$	(8) $32 \div 8$	(9) $108 \div 9$	(10) $22 \div 2$
(11) $44 \div 11$	(12) $48 \div 4$	(13) $30 \div 10$	(14) $90 \div 10$	(15) $100 \div 10$
(16) $0 \div 2$	(17) $9 \div 3$	(18) $70 \div 10$	(19) $0 \div 1$	(20) $55 \div 11$
(21) $33 \div 11$	(22) $0 \div 3$	(23) $20 \div 2$	(24) $20 \div 4$	(25) $5 \div 5$
(26) $40 \div 5$	(27) $44 \div 4$	(28) $35 \div 7$	(29) $0 \div 7$	(30) $18 \div 3$
(31) $24 \div 2$	(32) $2 \div 1$	(33) $16 \div 8$	(34) $4 \div 2$	(35) $6 \div 6$
(36) $8 \div 1$	(37) $110 \div 10$	(38) $7 \div 7$	(39) $21 \div 3$	(40) $80 \div 8$

90

Divide by 1-12

Time: __________

Score: _____/ 40

(1) 96 ÷ 12	(2) 99 ÷ 11	(3) 12 ÷ 2	(4) 120 ÷ 12	(5) 0 ÷ 11
(6) 24 ÷ 3	(7) 30 ÷ 6	(8) 32 ÷ 4	(9) 72 ÷ 12	(10) 48 ÷ 8
(11) 99 ÷ 9	(12) 24 ÷ 4	(13) 10 ÷ 2	(14) 90 ÷ 9	(15) 110 ÷ 11
(16) 8 ÷ 4	(17) 1 ÷ 1	(18) 12 ÷ 12	(19) 28 ÷ 7	(20) 63 ÷ 7
(21) 24 ÷ 6	(22) 0 ÷ 9	(23) 0 ÷ 6	(24) 4 ÷ 4	(25) 36 ÷ 6
(26) 14 ÷ 7	(27) 55 ÷ 5	(28) 24 ÷ 12	(29) 42 ÷ 7	(30) 20 ÷ 5
(31) 16 ÷ 4	(32) 28 ÷ 4	(33) 7 ÷ 1	(34) 22 ÷ 11	(35) 132 ÷ 12
(36) 24 ÷ 8	(37) 120 ÷ 10	(38) 12 ÷ 3	(39) 10 ÷ 5	(40) 12 ÷ 6

91

Divide by 1-12

Time: ____________

Score: ______/ 40

① 3 ÷ 1	② 15 ÷ 5	③ 0 ÷ 10	④ 40 ÷ 10	⑤ 16 ÷ 2
⑥ 4 ÷ 1	⑦ 27 ÷ 9	⑧ 50 ÷ 10	⑨ 12 ÷ 1	⑩ 11 ÷ 1
⑪ 60 ÷ 12	⑫ 6 ÷ 3	⑬ 48 ÷ 12	⑭ 45 ÷ 5	⑮ 18 ÷ 2
⑯ 36 ÷ 4	⑰ 54 ÷ 9	⑱ 81 ÷ 9	⑲ 50 ÷ 5	⑳ 0 ÷ 4
(21) 36 ÷ 9	(22) 10 ÷ 10	(23) 14 ÷ 2	(24) 36 ÷ 12	(25) 88 ÷ 8
(26) 88 ÷ 11	(27) 56 ÷ 7	(28) 54 ÷ 6	(29) 77 ÷ 11	(30) 60 ÷ 6
(31) 8 ÷ 8	(32) 45 ÷ 9	(33) 84 ÷ 12	(34) 5 ÷ 1	(35) 8 ÷ 2
(36) 21 ÷ 7	(37) 108 ÷ 12	(38) 80 ÷ 10	(39) 64 ÷ 8	(40) 30 ÷ 3

92

Divide by 1-12

Time: __________

Score: _____/ 40

(1) 42 ÷ 6	(2) 27 ÷ 3	(3) 40 ÷ 4	(4) 72 ÷ 6	(5) 15 ÷ 3
(6) 84 ÷ 7	(7) 144 ÷ 12	(8) 132 ÷ 11	(9) 70 ÷ 7	(10) 9 ÷ 1
(11) 66 ÷ 6	(12) 12 ÷ 4	(13) 96 ÷ 8	(14) 49 ÷ 7	(15) 72 ÷ 9
(16) 63 ÷ 9	(17) 77 ÷ 7	(18) 48 ÷ 6	(19) 2 ÷ 2	(20) 0 ÷ 5
(21) 40 ÷ 8	(22) 121 ÷ 11	(23) 3 ÷ 3	(24) 18 ÷ 9	(25) 0 ÷ 12
(26) 60 ÷ 5	(27) 25 ÷ 5	(28) 9 ÷ 9	(29) 48 ÷ 6	(30) 77 ÷ 7
(31) 24 ÷ 3	(32) 108 ÷ 12	(33) 0 ÷ 4	(34) 24 ÷ 12	(35) 36 ÷ 4
(36) 49 ÷ 7	(37) 45 ÷ 5	(38) 9 ÷ 9	(39) 16 ÷ 8	(40) 40 ÷ 4

93

Divide by 1-12

Time: ________

Score: _____/ 40

(1) 18 ÷ 9	(2) 2 ÷ 1	(3) 9 ÷ 3	(4) 54 ÷ 9	(5) 18 ÷ 6
(6) 0 ÷ 5	(7) 120 ÷ 10	(8) 0 ÷ 8	(9) 0 ÷ 2	(10) 12 ÷ 3
(11) 30 ÷ 5	(12) 18 ÷ 2	(13) 144 ÷ 12	(14) 77 ÷ 11	(15) 20 ÷ 4
(16) 36 ÷ 12	(17) 30 ÷ 6	(18) 80 ÷ 8	(19) 36 ÷ 3	(20) 15 ÷ 3
(21) 45 ÷ 9	(22) 16 ÷ 4	(23) 21 ÷ 3	(24) 66 ÷ 6	(25) 40 ÷ 5
(26) 48 ÷ 8	(27) 40 ÷ 8	(28) 84 ÷ 7	(29) 35 ÷ 5	(30) 70 ÷ 7
(31) 27 ÷ 9	(32) 33 ÷ 11	(33) 0 ÷ 11	(34) 14 ÷ 7	(35) 35 ÷ 7
(36) 30 ÷ 10	(37) 54 ÷ 6	(38) 48 ÷ 4	(39) 55 ÷ 5	(40) 28 ÷ 7

94

Divide by 1-12

Time: ____________

Score: _____/ 40

(1) 32 ÷ 4	(2) 12 ÷ 4	(3) 6 ÷ 6	(4) 48 ÷ 12	(5) 33 ÷ 3
(6) 60 ÷ 12	(7) 22 ÷ 2	(8) 6 ÷ 1	(9) 12 ÷ 6	(10) 18 ÷ 3
(11) 20 ÷ 10	(12) 0 ÷ 1	(13) 72 ÷ 6	(14) 55 ÷ 11	(15) 96 ÷ 12
(16) 90 ÷ 10	(17) 42 ÷ 7	(18) 36 ÷ 6	(19) 56 ÷ 8	(20) 4 ÷ 4
(21) 120 ÷ 12	(22) 11 ÷ 1	(23) 40 ÷ 10	(24) 8 ÷ 4	(25) 44 ÷ 11
(26) 16 ÷ 2	(27) 15 ÷ 5	(28) 6 ÷ 3	(29) 0 ÷ 7	(30) 42 ÷ 6
(31) 8 ÷ 2	(32) 5 ÷ 1	(33) 14 ÷ 2	(34) 90 ÷ 9	(35) 70 ÷ 10
(36) 21 ÷ 7	(37) 63 ÷ 9	(38) 110 ÷ 11	(39) 25 ÷ 5	(40) 0 ÷ 6

95

Divide by 1-12

Time: __________

Score: _____/ 40

(1) 99 ÷ 9	(2) 3 ÷ 3	(3) 60 ÷ 5	(4) 3 ÷ 1	(5) 24 ÷ 2
(6) 0 ÷ 9	(7) 88 ÷ 11	(8) 88 ÷ 8	(9) 44 ÷ 4	(10) 22 ÷ 11
(11) 81 ÷ 9	(12) 8 ÷ 8	(13) 121 ÷ 11	(14) 20 ÷ 2	(15) 24 ÷ 6
(16) 80 ÷ 10	(17) 5 ÷ 5	(18) 12 ÷ 12	(19) 56 ÷ 7	(20) 6 ÷ 2
(21) 2 ÷ 2	(22) 132 ÷ 12	(23) 36 ÷ 9	(24) 72 ÷ 12	(25) 7 ÷ 1
(26) 20 ÷ 5	(27) 8 ÷ 1	(28) 30 ÷ 3	(29) 4 ÷ 1	(30) 24 ÷ 8
(31) 1 ÷ 1	(32) 10 ÷ 2	(33) 12 ÷ 2	(34) 0 ÷ 3	(35) 10 ÷ 10
(36) 64 ÷ 8	(37) 50 ÷ 5	(38) 72 ÷ 8	(39) 72 ÷ 9	(40) 63 ÷ 7

96

Divide by 1-12

Time: ___________

Score: _____/ 40

1. 10 ÷ 5 = ____
2. 110 ÷ 10 = ____
3. 60 ÷ 10 = ____
4. 60 ÷ 6 = ____
5. 12 ÷ 1 = ____
6. 11 ÷ 11 = ____
7. 7 ÷ 7 = ____
8. 10 ÷ 1 = ____
9. 24 ÷ 4 = ____
10. 27 ÷ 3 = ____
11. 4 ÷ 2 = ____
12. 50 ÷ 10 = ____
13. 132 ÷ 11 = ____
14. 9 ÷ 1 = ____
15. 0 ÷ 12 = ____
16. 99 ÷ 11 = ____
17. 84 ÷ 12 = ____
18. 96 ÷ 8 = ____
19. 66 ÷ 11 = ____
20. 32 ÷ 8 = ____
21. 108 ÷ 9 = ____
22. 100 ÷ 10 = ____
23. 0 ÷ 10 = ____
24. 28 ÷ 4 = ____
25. 120 ÷ 10 = ____
26. 0 ÷ 1 = ____
27. 81 ÷ 9 = ____
28. 0 ÷ 2 = ____
29. 18 ÷ 2 = ____
30. 18 ÷ 6 = ____
31. 5 ÷ 5 = ____
32. 99 ÷ 9 = ____
33. 72 ÷ 9 = ____
34. 24 ÷ 8 = ____
35. 50 ÷ 5 = ____
36. 80 ÷ 10 = ____
37. 10 ÷ 1 = ____
38. 44 ÷ 11 = ____
39. 42 ÷ 6 = ____
40. 84 ÷ 7 = ____

97

Divide by 1-12

Time: ________

Score: _____/ 40

(1) 12 ÷ 12	(2) 36 ÷ 6	(3) 24 ÷ 6	(4) 16 ÷ 4	(5) 10 ÷ 2
(6) 54 ÷ 9	(7) 110 ÷ 10	(8) 40 ÷ 10	(9) 11 ÷ 11	(10) 42 ÷ 7
(11) 63 ÷ 9	(12) 72 ÷ 6	(13) 8 ÷ 4	(14) 5 ÷ 1	(15) 56 ÷ 7
(16) 2 ÷ 2	(17) 7 ÷ 1	(18) 66 ÷ 6	(19) 27 ÷ 3	(20) 30 ÷ 5
(21) 48 ÷ 12	(22) 33 ÷ 3	(23) 36 ÷ 9	(24) 48 ÷ 6	(25) 28 ÷ 4
(26) 77 ÷ 7	(27) 108 ÷ 12	(28) 8 ÷ 2	(29) 22 ÷ 2	(30) 4 ÷ 4
(31) 90 ÷ 9	(32) 55 ÷ 5	(33) 24 ÷ 2	(34) 144 ÷ 12	(35) 55 ÷ 11
(36) 12 ÷ 3	(37) 4 ÷ 2	(38) 8 ÷ 1	(39) 28 ÷ 7	(40) 6 ÷ 6

98

Divide by 1-12

Time: __________

Score: _____/ 40

(1) 0 ÷ 10	(2) 36 ÷ 12	(3) 77 ÷ 11	(4) 25 ÷ 5	(5) 45 ÷ 9
(6) 3 ÷ 3	(7) 32 ÷ 4	(8) 40 ÷ 8	(9) 44 ÷ 4	(10) 90 ÷ 10
(11) 12 ÷ 6	(12) 3 ÷ 1	(13) 0 ÷ 11	(14) 72 ÷ 12	(15) 110 ÷ 11
(16) 24 ÷ 3	(17) 18 ÷ 3	(18) 10 ÷ 10	(19) 0 ÷ 8	(20) 24 ÷ 12
(21) 8 ÷ 8	(22) 0 ÷ 3	(23) 70 ÷ 10	(24) 96 ÷ 8	(25) 36 ÷ 3
(26) 6 ÷ 3	(27) 60 ÷ 6	(28) 9 ÷ 3	(29) 121 ÷ 11	(30) 1 ÷ 1
(31) 72 ÷ 8	(32) 21 ÷ 7	(33) 15 ÷ 3	(34) 99 ÷ 11	(35) 56 ÷ 8
(36) 35 ÷ 7	(37) 88 ÷ 11	(38) 120 ÷ 12	(39) 60 ÷ 5	(40) 84 ÷ 12

99

Divide by 1-12

Time: ________

Score: _____/ 40

(1) 7 ÷ 7	(2) 63 ÷ 7	(3) 64 ÷ 8	(4) 88 ÷ 8	(5) 100 ÷ 10
(6) 20 ÷ 10	(7) 14 ÷ 7	(8) 6 ÷ 1	(9) 16 ÷ 8	(10) 18 ÷ 9
(11) 2 ÷ 1	(12) 12 ÷ 2	(13) 15 ÷ 5	(14) 27 ÷ 9	(15) 9 ÷ 9
(16) 80 ÷ 8	(17) 24 ÷ 4	(18) 0 ÷ 9	(19) 30 ÷ 3	(20) 22 ÷ 11
(21) 35 ÷ 5	(22) 50 ÷ 10	(23) 14 ÷ 2	(24) 0 ÷ 5	(25) 20 ÷ 2
(26) 48 ÷ 4	(27) 132 ÷ 11	(28) 54 ÷ 6	(29) 33 ÷ 11	(30) 132 ÷ 12
(31) 4 ÷ 1	(32) 12 ÷ 1	(33) 9 ÷ 1	(34) 60 ÷ 12	(35) 21 ÷ 3
(36) 36 ÷ 4	(37) 20 ÷ 5	(38) 0 ÷ 4	(39) 32 ÷ 8	(40) 48 ÷ 8

100

Divide by 1-12

Time: ________

Score: _____/ 40

(1) 0 ÷ 12	(2) 45 ÷ 5	(3) 11 ÷ 1	(4) 40 ÷ 5	(5) 66 ÷ 11
(6) 70 ÷ 7	(7) 0 ÷ 7	(8) 20 ÷ 4	(9) 0 ÷ 6	(10) 30 ÷ 6
(11) 12 ÷ 4	(12) 16 ÷ 2	(13) 30 ÷ 10	(14) 96 ÷ 12	(15) 40 ÷ 4
(16) 10 ÷ 5	(17) 6 ÷ 2	(18) 60 ÷ 10	(19) 49 ÷ 7	(20) 108 ÷ 9
(21) 18 ÷ 9	(22) 42 ÷ 6	(23) 24 ÷ 3	(24) 5 ÷ 1	(25) 12 ÷ 3
(26) 40 ÷ 5	(27) 24 ÷ 8	(28) 12 ÷ 4	(29) 36 ÷ 3	(30) 0 ÷ 5
(31) 90 ÷ 10	(32) 27 ÷ 3	(33) 27 ÷ 9	(34) 80 ÷ 8	(35) 56 ÷ 8
(36) 16 ÷ 8	(37) 12 ÷ 1	(38) 4 ÷ 4	(39) 36 ÷ 4	(40) 30 ÷ 10

Answer Key # 1

1. 64	2. 5	3. 1	4. 91	5. 74
6. 14	7. 7	8. 98	9. 46	10. 32
11. 45	12. 67	13. 38	14. 81	15. 27
16. 95	17. 62	18. 40	19. 19	20. 37
21. 71	22. 16	23. 83	24. 97	25. 86
26. 75	27. 42	28. 58	29. 77	30. 33
31. 84	32. 26	33. 18	34. 80	35. 3
36. 85	37. 20	38. 68	39. 22	40. 70

Answer Key # 2

1. 39	2. 88	3. 52	4. 57	5. 0
6. 92	7. 65	8. 61	9. 72	10. 29
11. 63	12. 24	13. 25	14. 93	15. 28
16. 21	17. 50	18. 89	19. 90	20. 87
21. 35	22. 9	23. 69	24. 82	25. 79
26. 76	27. 41	28. 56	29. 31	30. 96
31. 47	32. 30	33. 36	34. 59	35. 44
36. 94	37. 23	38. 4	39. 53	40. 73

Answer Key # 3

1. 78	2. 49	3. 10	4. 15	5. 99
6. 17	7. 13	8. 8	9. 2	10. 43
11. 55	12. 11	13. 100	14. 66	15. 51
16. 60	17. 12	18. 34	19. 48	20. 54
21. 6	22. 17	23. 38	24. 95	25. 66
26. 4	27. 42	28. 21	29. 80	30. 88
31. 34	32. 71	33. 93	34. 39	35. 30
36. 32	37. 36	38. 89	39. 48	40. 15

Answer Key # 4

1. 100	2. 9	3. 97	4. 50	5. 72
6. 44	7. 61	8. 77	9. 58	10. 98
11. 84	12. 35	13. 7	14. 26	15. 91
16. 25	17. 55	18. 86	19. 64	20. 0
21. 92	22. 78	23. 29	24. 63	25. 53
26. 52	27. 11	28. 74	29. 76	30. 14
31. 90	32. 83	33. 8	34. 12	35. 3
36. 31	37. 10	38. 23	39. 46	40. 47

Answer Key # 5

1. 6	2. 8	3. 3	4. 0	5. 5
6. 7	7. 11	8. 2	9. 9	10. 1
11. 10	12. 4	13. 12	14. 11	15. 2
16. 7	17. 5	18. 0	19. 9	20. 8
21. 3	22. 6	23. 4	24. 10	25. 12
26. 1	27. 8	28. 12	29. 0	30. 4
31. 10	32. 5	33. 6	34. 7	35. 2
36. 9	37. 1	38. 11	39. 3	40. 7

Answer Key # 6

1. 0	2. 8	3. 11	4. 5	5. 10
6. 2	7. 4	8. 6	9. 12	10. 9
11. 3	12. 1	13. 6	14. 1	15. 7
16. 5	17. 8	18. 4	19. 9	20. 0
21. 12	22. 10	23. 2	24. 3	25. 11
26. 6	27. 12	28. 2	29. 10	30. 8
31. 11	32. 3	33. 1	34. 9	35. 5
36. 4	37. 7	38. 0	39. 7	40. 10

Answer Key # 7

1. 8	2. 6	3. 4	4. 2	5. 11
6. 5	7. 9	8. 0	9. 12	10. 3
11. 1	12. 3	13. 4	14. 1	15. 10
16. 0	17. 7	18. 11	19. 2	20. 8
21. 9	22. 12	23. 5	24. 6	25. 0
26. 2	27. 8	28. 4	29. 12	30. 11
31. 9	32. 6	33. 1	34. 10	35. 3
36. 7	37. 5	38. 6	39. 4	40. 7

Answer Key # 8

1. 2	2. 12	3. 9	4. 3	5. 0
6. 8	7. 1	8. 11	9. 5	10. 10
11. 11	12. 10	13. 8	14. 4	15. 5
16. 9	17. 3	18. 6	19. 7	20. 12
21. 0	22. 2	23. 1	24. 5	25. 4
26. 3	27. 0	28. 10	29. 12	30. 11
31. 2	32. 6	33. 8	34. 1	35. 9
36. 7	37. 9	38. 0	39. 12	40. 5

Answer Key # 9

1. 7	2. 2	3. 5	4. 3	5. 8
6. 0	7. 1	8. 4	9. 9	10. 6
11. 5	12. 3	13. 2	14. 0	15. 7
16. 9	17. 8	18. 6	19. 4	20. 1
21. 0	22. 3	23. 8	24. 4	25. 5
26. 9	27. 6	28. 1	29. 7	30. 2
31. 1	32. 6	33. 5	34. 0	35. 9
36. 2	37. 8	38. 7	39. 3	40. 4

Answer Key # 10

1. 7	2. 4	3. 5	4. 8	5. 2
6. 1	7. 9	8. 6	9. 3	10. 0
11. 7	12. 5	13. 9	14. 3	15. 1
16. 4	17. 2	18. 8	19. 0	20. 6
21. 1	22. 7	23. 0	24. 6	25. 3
26. 4	27. 9	28. 2	29. 8	30. 5
31. 3	32. 1	33. 8	34. 7	35. 4
36. 5	37. 9	38. 0	39. 2	40. 6

Answer Key # 11

1. 2	2. 9	3. 3	4. 7	5. 5
6. 6	7. 0	8. 4	9. 8	10. 1
11. 5	12. 1	13. 8	14. 9	15. 7
16. 0	17. 2	18. 3	19. 6	20. 4
21. 5	22. 4	23. 7	24. 3	25. 0
26. 2	27. 6	28. 1	29. 9	30. 8
31. 3	32. 9	33. 7	34. 5	35. 1
36. 0	37. 4	38. 2	39. 6	40. 8

Answer Key # 12

1. 7	2. 9	3. 2	4. 3	5. 8
6. 5	7. 1	8. 6	9. 0	10. 4
11. 7	12. 4	13. 8	14. 3	15. 0
16. 5	17. 2	18. 1	19. 9	20. 6
21. 0	22. 4	23. 9	24. 2	25. 6
26. 7	27. 5	28. 8	29. 1	30. 3
31. 8	32. 3	33. 9	34. 6	35. 4
36. 2	37. 7	38. 1	39. 0	40. 5

Answer Key # 13

1. 1	2. 2	3. 0	4. 4	5. 3
6. 3	7. 8	8. 3	9. 6	10. 4
11. 0	12. 2	13. 7	14. 4	15. 7
16. 7	17. 9	18. 1	19. 0	20. 1
21. 5	22. 8	23. 6	24. 8	25. 5
26. 9	27. 2	28. 6	29. 5	30. 9
31. 7	32. 4	33. 8	34. 5	35. 9
36. 1	37. 0	38. 4	39. 2	40. 6

Answer Key # 14

1. 3	2. 3	3. 9	4. 0	5. 7
6. 4	7. 5	8. 8	9. 3	10. 1
11. 5	12. 6	13. 0	14. 9	15. 2
16. 1	17. 7	18. 6	19. 2	20. 8
21. 2	22. 6	23. 4	24. 3	25. 2
26. 1	27. 7	28. 8	29. 9	30. 3
31. 3	32. 5	33. 5	34. 1	35. 8
36. 9	37. 4	38. 7	39. 2	40. 1

Answer Key # 15

1. 5	2. 0	3. 0	4. 0	5. 6
6. 7	7. 8	8. 9	9. 6	10. 4
11. 6	12. 6	13. 9	14. 5	15. 4
16. 7	17. 2	18. 2	19. 8	20. 3
21. 1	22. 9	23. 1	24. 2	25. 0
26. 4	27. 3	28. 7	29. 4	30. 5
31. 7	32. 6	33. 0	34. 9	35. 1
36. 8	37. 3	38. 8	39. 0	40. 5

Answer Key # 16

1. 3	2. 2	3. 1	4. 3	5. 5
6. 6	7. 7	8. 8	9. 4	10. 9
11. 6	12. 1	13. 4	14. 9	15. 4
16. 8	17. 0	18. 2	19. 5	20. 8
21. 0	22. 0	23. 5	24. 3	25. 6
26. 2	27. 7	28. 9	29. 7	30. 1
31. 8	32. 2	33. 5	34. 6	35. 6
36. 1	37. 7	38. 1	39. 9	40. 6

Answer Key # 17

1. 4	2. 3	3. 9	4. 7	5. 2
6. 4	7. 3	8. 9	9. 7	10. 5
11. 0	12. 0	13. 2	14. 5	15. 8
16. 0	17. 3	18. 4	19. 8	20. 1
21. 7	22. 6	23. 0	24. 9	25. 3
26. 0	27. 3	28. 2	29. 3	30. 4
31. 9	32. 1	33. 5	34. 8	35. 0
36. 1	37. 4	38. 5	39. 6	40. 1

Answer Key # 18

1. 2	2. 7	3. 5	4. 8	5. 2
6. 7	7. 9	8. 4	9. 8	10. 6
11. 6	12. 5	13. 9	14. 6	15. 7
16. 1	17. 8	18. 9	19. 1	20. 4
21. 8	22. 6	23. 8	24. 4	25. 1
26. 0	27. 5	28. 3	29. 0	30. 3
31. 9	32. 7	33. 7	34. 2	35. 0
36. 3	37. 5	38. 2	39. 2	40. 4

Answer Key # 19

1. 3	2. 6	3. 5	4. 8	5. 7
6. 1	7. 2	8. 0	9. 2	10. 9
11. 4	12. 0	13. 8	14. 4	15. 4
16. 6	17. 9	18. 3	19. 5	20. 5
21. 2	22. 7	23. 0	24. 6	25. 3
26. 8	27. 7	28. 1	29. 1	30. 9
31. 7	32. 4	33. 2	34. 0	35. 9
36. 9	37. 8	38. 3	39. 0	40. 6

Answer Key # 20

1. 5	2. 5	3. 2	4. 6	5. 8
6. 6	7. 4	8. 1	9. 8	10. 7
11. 2	12. 0	13. 3	14. 3	15. 1
16. 1	17. 4	18. 5	19. 7	20. 9
21. 5	22. 8	23. 2	24. 4	25. 5
26. 0	27. 9	28. 1	29. 3	30. 0
31. 8	32. 0	33. 2	34. 3	35. 9
36. 6	37. 6	38. 7	39. 4	40. 3

Answer Key # 21

1. 3	2. 2	3. 7	4. 10	5. 11
6. 9	7. 5	8. 6	9. 12	10. 0
11. 1	12. 4	13. 8	14. 4	15. 7
16. 11	17. 10	18. 5	19. 3	20. 12
21. 1	22. 2	23. 9	24. 6	25. 8
26. 0	27. 12	28. 9	29. 0	30. 10
31. 3	32. 5	33. 7	34. 11	35. 1
36. 6	37. 4	38. 2	39. 8	40. 2

Answer Key # 22

1. 8	2. 12	3. 1	4. 3	5. 6
6. 5	7. 4	8. 7	9. 10	10. 9
11. 11	12. 0	13. 9	14. 8	15. 7
16. 2	17. 11	18. 6	19. 0	20. 5
21. 12	22. 10	23. 3	24. 4	25. 1
26. 12	27. 2	28. 8	29. 3	30. 6
31. 7	32. 9	33. 0	34. 4	35. 11
36. 5	37. 1	38. 10	39. 5	40. 2

Answer Key # 23

1. 0	2. 7	3. 10	4. 1	5. 9
6. 6	7. 11	8. 12	9. 4	10. 8
11. 3	12. 7	13. 2	14. 6	15. 9
16. 4	17. 11	18. 12	19. 10	20. 5
21. 8	22. 0	23. 1	24. 3	25. 6
26. 7	27. 9	28. 10	29. 5	30. 4
31. 3	32. 12	33. 11	34. 8	35. 0
36. 1	37. 2	38. 4	39. 0	40. 9

Answer Key # 24

1. 8	2. 6	3. 7	4. 5	5. 12
6. 11	7. 1	8. 3	9. 2	10. 10
11. 1	12. 3	13. 10	14. 11	15. 0
16. 9	17. 6	18. 2	19. 12	20. 8
21. 5	22. 7	23. 4	24. 10	25. 1
26. 5	27. 6	28. 0	29. 4	30. 9
31. 11	32. 7	33. 3	34. 2	35. 12
36. 8	37. 7	38. 8	39. 5	40. 9

Answer Key # 25

1. 1	2. 6	3. 5	4. 9	5. 0
6. 10	7. 4	8. 12	9. 3	10. 11
11. 8	12. 2	13. 7	14. 11	15. 3
16. 5	17. 4	18. 6	19. 10	20. 8
21. 7	22. 12	23. 1	24. 2	25. 9
26. 0	27. 6	28. 10	29. 1	30. 3
31. 8	32. 7	33. 11	34. 2	35. 9
36. 5	37. 0	38. 12	39. 4	40. 0

Answer Key # 26

1. 10	2. 2	3. 6	4. 12	5. 11
6. 3	7. 1	8. 4	9. 7	10. 9
11. 8	12. 5	13. 9	14. 10	15. 2
16. 7	17. 1	18. 3	19. 5	20. 4
21. 6	22. 0	23. 8	24. 11	25. 12
26. 2	27. 4	28. 12	29. 1	30. 6
31. 0	32. 11	33. 3	34. 9	35. 7
36. 8	37. 5	38. 10	39. 2	40. 9

Answer Key # 27

1. 5	2. 7	3. 11	4. 0	5. 6
6. 10	7. 8	8. 4	9. 12	10. 3
11. 1	12. 0	13. 9	14. 1	15. 12
16. 2	17. 5	18. 7	19. 11	20. 8
21. 10	22. 6	23. 3	24. 4	25. 5
26. 2	27. 9	28. 3	29. 7	30. 4
31. 12	32. 1	33. 10	34. 11	35. 6
36. 8	37. 0	38. 5	39. 0	40. 3

Answer Key # 28

1. 4	2. 1	3. 11	4. 12	5. 10
6. 2	7. 7	8. 9	9. 6	10. 8
11. 10	12. 4	13. 5	14. 3	15. 6
16. 11	17. 1	18. 0	19. 2	20. 12
21. 7	22. 8	23. 9	24. 7	25. 12
26. 4	27. 10	28. 0	29. 1	30. 2
31. 3	32. 9	33. 8	34. 6	35. 11
36. 5	37. 6	38. 7	39. 9	40. 8

Answer Key # 29

1. 7	2. 0	3. 9	4. 11	5. 12
6. 10	7. 5	8. 4	9. 3	10. 8
11. 2	12. 1	13. 6	14. 11	15. 6
16. 5	17. 10	18. 2	19. 9	20. 3
21. 1	22. 0	23. 4	24. 7	25. 8
26. 12	27. 7	28. 12	29. 8	30. 10
31. 3	32. 0	33. 1	34. 4	35. 2
36. 5	37. 6	38. 11	39. 9	40. 1

Answer Key # 30

1. 2	2. 5	3. 12	4. 10	5. 6
6. 0	7. 4	8. 11	9. 8	10. 7
11. 3	12. 9	13. 3	14. 7	15. 4
16. 6	17. 9	18. 10	19. 2	20. 12
21. 8	22. 5	23. 1	24. 11	25. 0
26. 6	27. 10	28. 8	29. 9	30. 4
31. 12	32. 3	33. 0	34. 7	35. 5
36. 2	37. 1	38. 11	39. 8	40. 11

Answer Key # 31

1. 6	2. 0	3. 4	4. 7	5. 1
6. 3	7. 10	8. 12	9. 9	10. 2
11. 5	12. 11	13. 1	14. 5	15. 3
16. 0	17. 9	18. 7	19. 2	20. 6
21. 10	22. 12	23. 4	24. 8	25. 5
26. 11	27. 8	28. 0	29. 4	30. 12
31. 2	32. 7	33. 3	34. 6	35. 1
36. 9	37. 10	38. 5	39. 10	40. 2

Answer Key # 32

1. 6	2. 12	3. 7	4. 3	5. 8
6. 4	7. 9	8. 0	9. 1	10. 11
11. 7	12. 8	13. 12	14. 5	15. 3
16. 1	17. 11	18. 0	19. 4	20. 9
21. 6	22. 2	23. 10	24. 0	25. 11
26. 1	27. 2	28. 8	29. 6	30. 9
31. 7	32. 5	33. 12	34. 4	35. 3
36. 10	37. 1	38. 9	39. 6	40. 0

Answer Key # 33

1. 2	2. 10	3. 11	4. 1	5. 3
6. 1	7. 8	8. 11	9. 11	10. 7
11. 6	12. 8	13. 2	14. 9	15. 9
16. 4	17. 10	18. 5	19. 5	20. 2
21. 1	22. 10	23. 9	24. 7	25. 12
26. 4	27. 6	28. 0	29. 0	30. 12
31. 0	32. 6	33. 4	34. 3	35. 3
36. 8	37. 7	38. 12	39. 5	40. 9

Answer Key # 34

1. 4	2. 1	3. 6	4. 11	5. 9
6. 6	7. 5	8. 12	9. 9	10. 11
11. 10	12. 11	13. 12	14. 5	15. 7
16. 12	17. 0	18. 2	19. 1	20. 4
21. 1	22. 0	23. 2	24. 8	25. 6
26. 10	27. 3	28. 4	29. 7	30. 3
31. 3	32. 7	33. 8	34. 2	35. 5
36. 10	37. 8	38. 0	39. 12	40. 1

Answer Key # 35

1. 8	2. 2	3. 3	4. 6	5. 1
6. 0	7. 2	8. 10	9. 5	10. 2
11. 12	12. 8	13. 0	14. 11	15. 8
16. 6	17. 7	18. 6	19. 11	20. 3
21. 1	22. 10	23. 5	24. 11	25. 7
26. 7	27. 5	28. 9	29. 0	30. 9
31. 10	32. 3	33. 4	34. 9	35. 4
36. 12	37. 4	38. 5	39. 10	40. 3

Answer Key # 36

1. 7	2. 2	3. 0	4. 5	5. 7
6. 3	7. 2	8. 10	9. 12	10. 0
11. 0	12. 9	13. 6	14. 1	15. 5
16. 4	17. 12	18. 11	19. 8	20. 7
21. 3	22. 9	23. 4	24. 12	25. 6
26. 1	27. 10	28. 1	29. 6	30. 9
31. 8	32. 4	33. 11	34. 11	35. 2
36. 8	37. 6	38. 12	39. 2	40. 1

Answer Key # 37

1. 3	2. 11	3. 7	4. 2	5. 1
6. 8	7. 3	8. 10	9. 10	10. 9
11. 5	12. 11	13. 0	14. 0	15. 10
16. 12	17. 8	18. 12	19. 5	20. 4
21. 9	22. 9	23. 4	24. 11	25. 0
26. 6	27. 3	28. 4	29. 1	30. 2
31. 5	32. 8	33. 7	34. 7	35. 6
36. 3	37. 3	38. 2	39. 10	40. 5

Answer Key # 38

1. 1	2. 0	3. 4	4. 6	5. 12
6. 7	7. 6	8. 12	9. 1	10. 1
11. 11	12. 9	13. 9	14. 4	15. 2
16. 6	17. 0	18. 12	19. 0	20. 10
21. 7	22. 4	23. 8	24. 11	25. 7
26. 11	27. 8	28. 3	29. 2	30. 8
31. 5	32. 9	33. 5	34. 10	35. 4
36. 1	37. 5	38. 5	39. 6	40. 6

Answer Key # 39

1. 12	2. 9	3. 2	4. 8	5. 1
6. 2	7. 0	8. 10	9. 5	10. 0
11. 12	12. 10	13. 3	14. 3	15. 10
16. 9	17. 7	18. 11	19. 1	20. 8
21. 4	22. 4	23. 12	24. 11	25. 3
26. 11	27. 6	28. 8	29. 2	30. 7
31. 9	32. 0	33. 7	34. 7	35. 2
36. 6	37. 1	38. 9	39. 5	40. 0

Answer Key # 40

1. 6	2. 6	3. 9	4. 10	5. 2
6. 4	7. 4	8. 5	9. 3	10. 8
11. 8	12. 4	13. 11	14. 7	15. 10
16. 5	17. 12	18. 0	19. 12	20. 3
21. 1	22. 0	23. 2	24. 1	25. 10
26. 11	27. 3	28. 12	29. 8	30. 7
31. 11	32. 9	33. 1	34. 2	35. 5
36. 0	37. 9	38. 9	39. 1	40. 7

Answer Key # 41

1. 7	2. 4	3. 12	4. 2	5. 6
6. 5	7. 9	8. 8	9. 0	10. 11
11. 1	12. 3	13. 10	14. 11	15. 10
16. 6	17. 3	18. 0	19. 8	20. 2
21. 5	22. 9	23. 4	24. 7	25. 1
26. 12	27. 5	28. 9	29. 12	30. 2
31. 0	32. 10	33. 11	34. 8	35. 6
36. 4	37. 1	38. 7	39. 3	40. 8

Answer Key # 42

1. 12	2. 5	3. 6	4. 0	5. 9
6. 2	7. 1	8. 3	9. 11	10. 10
11. 4	12. 7	13. 9	14. 7	15. 11
16. 1	17. 0	18. 6	19. 5	20. 3
21. 2	22. 10	23. 8	24. 12	25. 4
26. 2	27. 10	28. 7	29. 1	30. 6
31. 9	32. 12	33. 0	34. 5	35. 3
36. 8	37. 4	38. 11	39. 7	40. 12

Answer Key # 43

1. 5	2. 6	3. 11	4. 4	5. 1
6. 2	7. 9	8. 10	9. 0	10. 3
11. 8	12. 11	13. 3	14. 10	15. 6
16. 8	17. 5	18. 12	19. 0	20. 2
21. 4	22. 7	23. 1	24. 9	25. 0
26. 9	27. 2	28. 4	29. 11	30. 1
31. 5	32. 10	33. 3	34. 8	35. 6
36. 7	37. 12	38. 2	39. 10	40. 12

Answer Key # 44

1. 0	2. 7	3. 5	4. 8	5. 11
6. 3	7. 1	8. 9	9. 6	10. 4
11. 6	12. 12	13. 0	14. 5	15. 8
16. 10	17. 2	18. 7	19. 4	20. 1
21. 3	22. 11	23. 9	24. 0	25. 12
26. 11	27. 3	28. 2	29. 7	30. 5
31. 6	32. 4	33. 9	34. 8	35. 1
36. 10	37. 5	38. 11	39. 0	40. 10

Answer Key # 45

1. 2	2. 7	3. 10	4. 9	5. 5
6. 8	7. 1	8. 12	9. 11	10. 6
11. 3	12. 0	13. 4	14. 10	15. 4
16. 11	17. 9	18. 3	19. 5	20. 0
21. 8	22. 2	23. 6	24. 1	25. 12
26. 7	27. 11	28. 7	29. 10	30. 1
31. 5	32. 8	33. 3	34. 2	35. 12
36. 0	37. 9	38. 4	39. 6	40. 5

Answer Key # 46

1. 6	2. 8	3. 0	4. 7	5. 10
6. 1	7. 9	8. 12	9. 4	10. 2
11. 3	12. 11	13. 6	14. 8	15. 10
16. 7	17. 11	18. 3	19. 1	20. 9
21. 12	22. 5	23. 2	24. 4	25. 0
26. 3	27. 10	28. 9	29. 8	30. 12
31. 0	32. 2	33. 7	34. 6	35. 4
36. 5	37. 11	38. 1	39. 5	40. 9

Answer Key # 47

1. 8	2. 2	3. 1	4. 0	5. 11
6. 10	7. 4	8. 12	9. 6	10. 7
11. 3	12. 0	13. 7	14. 5	15. 6
16. 9	17. 3	18. 10	19. 1	20. 8
21. 12	22. 2	23. 11	24. 4	25. 9
26. 3	27. 7	28. 1	29. 10	30. 11
31. 8	32. 12	33. 2	34. 6	35. 4
36. 5	37. 0	38. 12	39. 8	40. 7

Answer Key # 48

1. 11	2. 10	3. 4	4. 9	5. 5
6. 2	7. 0	8. 6	9. 1	10. 3
11. 12	12. 5	13. 11	14. 6	15. 8
16. 0	17. 9	18. 10	19. 7	20. 2
21. 4	22. 1	23. 3	24. 5	25. 3
26. 4	27. 2	28. 11	29. 1	30. 6
31. 12	32. 10	33. 8	34. 7	35. 9
36. 0	37. 12	38. 10	39. 1	40. 11

Answer Key # 49

1. 2	2. 7	3. 6	4. 4	5. 1
6. 11	7. 0	8. 8	9. 9	10. 3
11. 5	12. 10	13. 12	14. 9	15. 3
16. 6	17. 0	18. 10	19. 12	20. 8
21. 4	22. 11	23. 7	24. 5	25. 2
26. 1	27. 6	28. 8	29. 3	30. 0
31. 2	32. 9	33. 11	34. 7	35. 5
36. 1	37. 12	38. 4	39. 10	40. 11

Answer Key # 50

1. 8	2. 6	3. 9	4. 3	5. 5
6. 12	7. 0	8. 1	9. 7	10. 4
11. 10	12. 2	13. 5	14. 2	15. 11
16. 12	17. 9	18. 10	19. 6	20. 4
21. 0	22. 1	23. 7	24. 3	25. 8
26. 2	27. 0	28. 8	29. 3	30. 12
31. 4	32. 10	33. 11	34. 5	35. 1
36. 9	37. 7	38. 6	39. 2	40. 4

Answer Key # 51

1. 11	2. 8	3. 7	4. 10	5. 5
6. 3	7. 9	8. 1	9. 0	10. 12
11. 6	12. 10	13. 6	14. 5	15. 0
16. 7	17. 4	18. 12	19. 3	20. 8
21. 1	22. 9	23. 2	24. 11	25. 1
26. 11	27. 5	28. 10	29. 0	30. 12
31. 2	32. 8	33. 3	34. 6	35. 4
36. 7	37. 9	38. 2	39. 7	40. 0

Answer Key # 52

1. 10	2. 8	3. 11	4. 5	5. 12
6. 4	7. 1	8. 9	9. 6	10. 3
11. 5	12. 8	13. 10	14. 6	15. 11
16. 3	17. 9	18. 1	19. 12	20. 4
21. 0	22. 2	23. 7	24. 10	25. 2
26. 7	27. 9	28. 1	29. 12	30. 0
31. 8	32. 5	33. 4	34. 3	35. 11
36. 6	37. 10	38. 9	39. 1	40. 11

Answer Key # 53

1. 2	2. 10	3. 1	4. 6	5. 8
6. 3	7. 10	8. 9	9. 11	10. 4
11. 7	12. 1	13. 12	14. 12	15. 8
16. 10	17. 7	18. 9	19. 9	20. 5
21. 4	22. 12	23. 1	24. 8	25. 0
26. 5	27. 3	28. 2	29. 6	30. 5
31. 11	32. 0	33. 11	34. 3	35. 6
36. 4	37. 7	38. 2	39. 0	40. 10

Answer Key # 54

1. 0	2. 1	3. 3	4. 9	5. 12
6. 12	7. 5	8. 2	9. 9	10. 8
11. 12	12. 10	13. 2	14. 7	15. 5
16. 7	17. 3	18. 2	19. 1	20. 5
21. 8	22. 11	23. 6	24. 9	25. 0
26. 0	27. 11	28. 4	29. 4	30. 8
31. 11	32. 1	33. 6	34. 10	35. 7
36. 3	37. 6	38. 4	39. 11	40. 4

Answer Key # 55

1. 1	2. 6	3. 5	4. 11	5. 1
6. 4	7. 11	8. 7	9. 5	10. 5
11. 1	12. 6	13. 3	14. 7	15. 10
16. 3	17. 0	18. 8	19. 10	20. 12
21. 9	22. 4	23. 2	24. 8	25. 12
26. 3	27. 2	28. 9	29. 12	30. 8
31. 9	32. 6	33. 0	34. 2	35. 0
36. 7	37. 10	38. 10	39. 7	40. 4

Answer Key # 56

1. 0	2. 3	3. 0	4. 5	5. 9
6. 4	7. 5	8. 12	9. 10	10. 8
11. 1	12. 2	13. 8	14. 1	15. 3
16. 4	17. 2	18. 11	19. 9	20. 8
21. 9	22. 11	23. 3	24. 7	25. 1
26. 10	27. 5	28. 7	29. 2	30. 6
31. 11	32. 6	33. 12	34. 0	35. 12
36. 6	37. 0	38. 7	39. 3	40. 2

Answer Key # 57

1. 5	2. 3	3. 7	4. 10	5. 0
6. 5	7. 8	8. 9	9. 4	10. 10
11. 9	12. 7	13. 11	14. 3	15. 2
16. 11	17. 4	18. 2	19. 1	20. 12
21. 6	22. 12	23. 5	24. 8	25. 12
26. 6	27. 8	28. 0	29. 9	30. 1
31. 4	32. 1	33. 10	34. 11	35. 6
36. 9	37. 12	38. 1	39. 12	40. 2

Answer Key # 58

1. 2	2. 7	3. 4	4. 6	5. 8
6. 2	7. 12	8. 10	9. 1	10. 7
11. 11	12. 10	13. 1	14. 0	15. 11
16. 3	17. 11	18. 9	19. 5	20. 9
21. 3	22. 4	23. 8	24. 7	25. 6
26. 5	27. 0	28. 3	29. 0	30. 5
31. 6	32. 10	33. 8	34. 4	35. 6
36. 0	37. 11	38. 4	39. 5	40. 0

Answer Key # 59

1. 6	2. 2	3. 8	4. 10	5. 10
6. 12	7. 1	8. 9	9. 12	10. 3
11. 10	12. 7	13. 9	14. 11	15. 7
16. 1	17. 2	18. 2	19. 5	20. 0
21. 3	22. 6	23. 8	24. 5	25. 11
26. 1	27. 4	28. 8	29. 4	30. 12
31. 3	32. 9	33. 7	34. 2	35. 8
36. 2	37. 8	38. 5	39. 10	40. 6

Answer Key # 60

1. 10	2. 2	3. 3	4. 0	5. 1
6. 7	7. 1	8. 12	9. 9	10. 0
11. 3	12. 3	13. 12	14. 10	15. 5
16. 6	17. 11	18. 1	19. 11	20. 4
21. 4	22. 7	23. 11	24. 9	25. 5
26. 12	27. 8	28. 0	29. 6	30. 7
31. 9	32. 4	33. 1	34. 0	35. 3
36. 1	37. 10	38. 2	39. 6	40. 8

Answer Key # 61

1. 8	2. 4	3. 0	4. 9	5. 12
6. 10	7. 5	8. 1	9. 2	10. 11
11. 7	12. 3	13. 6	14. 1	15. 8
16. 7	17. 9	18. 6	19. 0	20. 3
21. 2	22. 11	23. 10	24. 12	25. 4
26. 5	27. 12	28. 7	29. 6	30. 9
31. 0	32. 3	33. 8	34. 2	35. 10
36. 5	37. 1	38. 4	39. 11	40. 12

Answer Key # 62

1. 10	2. 6	3. 5	4. 0	5. 8
6. 7	7. 9	8. 11	9. 2	10. 1
11. 3	12. 4	13. 5	14. 6	15. 1
16. 7	17. 0	18. 4	19. 12	20. 10
21. 8	22. 2	23. 11	24. 3	25. 9
26. 10	27. 4	28. 7	29. 8	30. 5
31. 1	32. 3	33. 0	34. 11	35. 12
36. 2	37. 6	38. 9	39. 8	40. 2

Answer Key # 63

1. 1	2. 3	3. 0	4. 12	5. 10
6. 9	7. 6	8. 5	9. 4	10. 7
11. 11	12. 0	13. 4	14. 5	15. 8
16. 2	17. 11	18. 10	19. 12	20. 7
21. 6	22. 3	23. 1	24. 9	25. 8
26. 9	27. 6	28. 4	29. 1	30. 12
31. 5	32. 10	33. 3	34. 7	35. 2
36. 11	37. 0	38. 10	39. 7	40. 11

Answer Key # 64

1. 2	2. 5	3. 9	4. 3	5. 12
6. 8	7. 0	8. 6	9. 4	10. 1
11. 2	12. 7	13. 1	14. 8	15. 6
16. 0	17. 11	18. 4	19. 10	20. 12
21. 5	22. 9	23. 3	24. 7	25. 3
26. 10	27. 11	28. 2	29. 5	30. 1
31. 8	32. 12	33. 6	34. 9	35. 0
36. 4	37. 9	38. 3	39. 11	40. 2

Answer Key # 65

1. 5	2. 9	3. 1	4. 3	5. 7
6. 6	7. 4	8. 10	9. 8	10. 12
11. 11	12. 0	13. 2	14. 12	15. 11
16. 4	17. 5	18. 10	19. 9	20. 1
21. 8	22. 3	23. 6	24. 7	25. 2
26. 0	27. 3	28. 4	29. 6	30. 2
31. 0	32. 1	33. 12	34. 5	35. 8
36. 9	37. 10	38. 7	39. 11	40. 4

Answer Key # 66

1. 11	2. 3	3. 8	4. 10	5. 1
6. 9	7. 7	8. 0	9. 5	10. 6
11. 12	12. 2	13. 5	14. 12	15. 6
16. 10	17. 8	18. 7	19. 1	20. 11
21. 0	22. 9	23. 2	24. 4	25. 3
26. 12	27. 10	28. 4	29. 5	30. 3
31. 1	32. 2	33. 7	34. 0	35. 6
36. 11	37. 8	38. 9	39. 5	40. 12

Answer Key # 67

1. 9	2. 6	3. 11	4. 4	5. 1
6. 0	7. 7	8. 2	9. 8	10. 10
11. 3	12. 6	13. 2	14. 0	15. 12
16. 9	17. 5	18. 4	19. 1	20. 3
21. 11	22. 8	23. 7	24. 10	25. 7
26. 6	27. 9	28. 8	29. 12	30. 2
31. 3	32. 11	33. 5	34. 10	35. 1
36. 0	37. 4	38. 11	39. 9	40. 0

Answer Key # 68

1. 3	2. 2	3. 1	4. 4	5. 8
6. 7	7. 6	8. 5	9. 12	10. 10
11. 6	12. 0	13. 10	14. 1	15. 12
16. 8	17. 4	18. 9	19. 5	20. 7
21. 3	22. 11	23. 2	24. 5	25. 8
26. 1	27. 7	28. 6	29. 0	30. 11
31. 12	32. 3	33. 2	34. 10	35. 4
36. 9	37. 11	38. 5	39. 4	40. 12

Answer Key # 69

1. 8	2. 7	3. 2	4. 0	5. 10
6. 12	7. 4	8. 5	9. 9	10. 3
11. 1	12. 11	13. 6	14. 0	15. 7
16. 12	17. 3	18. 6	19. 10	20. 9
21. 4	22. 1	23. 2	24. 5	25. 11
26. 8	27. 3	28. 11	29. 7	30. 4
31. 0	32. 10	33. 12	34. 8	35. 6
36. 1	37. 2	38. 9	39. 5	40. 9

Answer Key # 70

1. 11	2. 4	3. 6	4. 10	5. 2
6. 12	7. 8	8. 1	9. 0	10. 7
11. 3	12. 5	13. 6	14. 5	15. 3
16. 7	17. 12	18. 10	19. 11	20. 2
21. 0	22. 4	23. 1	24. 9	25. 8
26. 1	27. 4	28. 8	29. 2	30. 9
31. 6	32. 10	33. 11	34. 12	35. 0
36. 7	37. 3	38. 5	39. 10	40. 3

Answer Key # 71

1. 5	2. 11	3. 9	4. 6	5. 2
6. 4	7. 1	8. 7	9. 8	10. 12
11. 0	12. 12	13. 2	14. 11	15. 9
16. 10	17. 1	18. 6	19. 3	20. 8
21. 4	22. 5	23. 0	24. 7	25. 6
26. 0	27. 2	28. 1	29. 11	30. 12
31. 10	32. 9	33. 5	34. 3	35. 8
36. 4	37. 7	38. 8	39. 0	40. 6

Answer Key # 72

1. 4	2. 10	3. 12	4. 5	5. 2
6. 3	7. 9	8. 11	9. 7	10. 1
11. 9	12. 5	13. 0	14. 10	15. 6
16. 3	17. 7	18. 2	19. 8	20. 1
21. 4	22. 11	23. 12	24. 0	25. 1
26. 5	27. 7	28. 11	29. 4	30. 8
31. 12	32. 2	33. 9	34. 10	35. 3
36. 6	37. 11	38. 3	39. 6	40. 8

Answer Key # 73

1. 2	2. 10	3. 5	4. 8	5. 6
6. 1	7. 1	8. 5	9. 9	10. 3
11. 0	12. 6	13. 9	14. 12	15. 5
16. 8	17. 7	18. 1	19. 4	20. 10
21. 9	22. 12	23. 7	24. 0	25. 2
26. 11	27. 11	28. 10	29. 8	30. 6
31. 3	32. 4	33. 12	34. 11	35. 2
36. 0	37. 7	38. 3	39. 4	40. 6

Answer Key # 74

1. 12	2. 10	3. 11	4. 10	5. 5
6. 12	7. 7	8. 12	9. 6	10. 0
11. 9	12. 5	13. 2	14. 2	15. 0
16. 10	17. 11	18. 0	19. 9	20. 3
21. 4	22. 4	23. 5	24. 11	25. 1
26. 3	27. 8	28. 8	29. 7	30. 9
31. 6	32. 1	33. 1	34. 7	35. 4
36. 3	37. 8	38. 2	39. 0	40. 2

Answer Key # 75

1. 9	2. 12	3. 9	4. 12	5. 10
6. 3	7. 5	8. 5	9. 7	10. 9
11. 0	12. 1	13. 6	14. 0	15. 7
16. 8	17. 4	18. 11	19. 10	20. 2
21. 8	22. 11	23. 3	24. 6	25. 4
26. 5	27. 12	28. 2	29. 4	30. 7
31. 6	32. 1	33. 11	34. 1	35. 10
36. 3	37. 8	38. 4	39. 4	40. 8

Answer Key # 76

1. 2	2. 6	3. 10	4. 2	5. 1
6. 3	7. 12	8. 6	9. 6	10. 9
11. 7	12. 9	13. 5	14. 12	15. 10
16. 12	17. 8	18. 2	19. 1	20. 7
21. 3	22. 0	23. 4	24. 0	25. 3
26. 8	27. 9	28. 5	29. 1	30. 10
31. 7	32. 11	33. 5	34. 0	35. 11
36. 11	37. 4	38. 10	39. 9	40. 11

Answer Key # 77

1. 12	2. 4	3. 3	4. 3	5. 1
6. 3	7. 5	8. 2	9. 12	10. 6
11. 12	12. 2	13. 10	14. 9	15. 11
16. 9	17. 6	18. 0	19. 7	20. 7
21. 2	22. 0	23. 1	24. 6	25. 4
26. 1	27. 5	28. 8	29. 7	30. 11
31. 10	32. 8	33. 5	34. 8	35. 0
36. 0	37. 5	38. 2	39. 1	40. 7

Answer Key # 78

1. 2	2. 1	3. 12	4. 3	5. 11
6. 3	7. 11	8. 10	9. 4	10. 0
11. 12	12. 8	13. 9	14. 6	15. 10
16. 8	17. 7	18. 3	19. 4	20. 2
21. 5	22. 4	23. 5	24. 1	25. 11
26. 8	27. 7	28. 9	29. 10	30. 6
31. 6	32. 0	33. 9	34. 12	35. 0
36. 2	37. 10	38. 1	39. 1	40. 12

Answer Key # 79

1. 8	2. 2	3. 7	4. 0	5. 11
6. 7	7. 12	8. 0	9. 3	10. 10
11. 5	12. 4	13. 1	14. 11	15. 9
16. 5	17. 8	18. 3	19. 6	20. 9
21. 4	22. 11	23. 3	24. 2	25. 12
26. 10	27. 9	28. 4	29. 8	30. 6
31. 5	32. 7	33. 6	34. 9	35. 1
36. 1	37. 6	38. 11	39. 6	40. 4

Answer Key # 80

1. 7	2. 4	3. 5	4. 5	5. 10
6. 3	7. 9	8. 7	9. 10	10. 0
11. 3	12. 10	13. 1	14. 8	15. 11
16. 2	17. 9	18. 6	19. 2	20. 12
21. 12	22. 0	23. 0	24. 8	25. 5
26. 8	27. 4	28. 11	29. 7	30. 3
31. 2	32. 12	33. 12	34. 10	35. 12
36. 10	37. 12	38. 1	39. 3	40. 9

Answer Key # 81

1. 6	2. 11	3. 5	4. 10	5. 1
6. 10	7. 3	8. 6	9. 1	10. 2
11. 7	12. 2	13. 10	14. 8	15. 6
16. 5	17. 2	18. 11	19. 8	20. 12
21. 4	22. 5	23. 5	24. 4	25. 2
26. 1	27. 2	28. 7	29. 7	30. 5
31. 5	32. 5	33. 6	34. 12	35. 10
36. 11	37. 8	38. 5	39. 12	40. 12

Answer Key # 82

1. 0	2. 7	3. 10	4. 3	5. 1
6. 0	7. 0	8. 6	9. 7	10. 4
11. 0	12. 9	13. 7	14. 0	15. 11
16. 3	17. 11	18. 4	19. 9	20. 12
21. 9	22. 10	23. 4	24. 0	25. 6
26. 2	27. 12	28. 10	29. 0	30. 1
31. 8	32. 6	33. 6	34. 7	35. 12
36. 2	37. 3	38. 3	39. 6	40. 3

Answer Key # 83

1. 8	2. 0	3. 7	4. 5	5. 1
6. 8	7. 3	8. 7	9. 0	10. 1
11. 2	12. 12	13. 11	14. 4	15. 6
16. 8	17. 9	18. 10	19. 4	20. 3
21. 11	22. 3	23. 3	24. 1	25. 0
26. 1	27. 9	28. 7	29. 12	30. 9
31. 8	32. 4	33. 1	34. 5	35. 10
36. 4	37. 12	38. 2	39. 8	40. 10

Answer Key # 84

1. 1	2. 6	3. 2	4. 4	5. 3
6. 11	7. 1	8. 6	9. 9	10. 0
11. 9	12. 7	13. 8	14. 9	15. 11
16. 5	17. 8	18. 10	19. 0	20. 12
21. 10	22. 11	23. 3	24. 8	25. 5
26. 9	27. 11	28. 2	29. 4	30. 9
31. 2	32. 12	33. 11	34. 4	35. 9
36. 7	37. 7	38. 3	39. 0	40. 7

Answer Key # 85

1. 11	2. 8	3. 1	4. 12	5. 5
6. 0	7. 11	8. 11	9. 2	10. 11
11. 7	12. 6	13. 1	14. 3	15. 9
16. 0	17. 2	18. 6	19. 9	20. 12
21. 3	22. 9	23. 4	24. 4	25. 5
26. 0	27. 9	28. 12	29. 9	30. 6
31. 4	32. 8	33. 6	34. 1	35. 4
36. 0	37. 10	38. 11	39. 11	40. 1

Answer Key # 86

1. 4	2. 5	3. 1	4. 2	5. 12
6. 9	7. 10	8. 7	9. 10	10. 3
11. 4	12. 0	13. 11	14. 0	15. 3
16. 12	17. 7	18. 3	19. 10	20. 8
21. 8	22. 6	23. 5	24. 4	25. 5
26. 2	27. 0	28. 12	29. 7	30. 10
31. 5	32. 7	33. 0	34. 10	35. 5
36. 5	37. 12	38. 1	39. 8	40. 7

Answer Key # 87

1. 8	2. 9	3. 10	4. 2	5. 10
6. 10	7. 3	8. 10	9. 3	10. 4
11. 12	12. 6	13. 5	14. 8	15. 11
16. 12	17. 3	18. 6	19. 0	20. 3
21. 4	22. 2	23. 5	24. 5	25. 7
26. 2	27. 8	28. 9	29. 0	30. 3
31. 2	32. 6	33. 2	34. 1	35. 11
36. 12	37. 10	38. 5	39. 8	40. 6

Answer Key # 88

1. 2	2. 2	3. 11	4. 9	5. 6
6. 10	7. 1	8. 9	9. 11	10. 4
11. 4	12. 6	13. 12	14. 8	15. 7
16. 6	17. 2	18. 12	19. 9	20. 1
21. 4	22. 1	23. 1	24. 8	25. 11
26. 9	27. 7	28. 3	29. 7	30. 0
31. 1	32. 8	33. 10	34. 3	35. 2
36. 0	37. 1	38. 6	39. 9	40. 12

Answer Key # 89

1. 11	2. 3	3. 7	4. 6	5. 7
6. 6	7. 6	8. 4	9. 12	10. 11
11. 4	12. 12	13. 3	14. 9	15. 10
16. 0	17. 3	18. 7	19. 0	20. 5
21. 3	22. 0	23. 10	24. 5	25. 1
26. 8	27. 11	28. 5	29. 0	30. 6
31. 12	32. 2	33. 2	34. 2	35. 1
36. 8	37. 11	38. 1	39. 7	40. 10

Answer Key # 90

1. 8	2. 9	3. 6	4. 10	5. 0
6. 8	7. 5	8. 8	9. 6	10. 6
11. 11	12. 6	13. 5	14. 10	15. 10
16. 2	17. 1	18. 1	19. 4	20. 9
21. 4	22. 0	23. 0	24. 1	25. 6
26. 2	27. 11	28. 2	29. 6	30. 4
31. 4	32. 7	33. 7	34. 2	35. 11
36. 3	37. 12	38. 4	39. 2	40. 2

Answer Key # 91

1. 3	2. 3	3. 0	4. 4	5. 8
6. 4	7. 3	8. 5	9. 12	10. 11
11. 5	12. 2	13. 4	14. 9	15. 9
16. 9	17. 6	18. 9	19. 10	20. 0
21. 4	22. 1	23. 7	24. 3	25. 11
26. 8	27. 8	28. 9	29. 7	30. 10
31. 1	32. 5	33. 7	34. 5	35. 4
36. 3	37. 9	38. 8	39. 8	40. 10

Answer Key # 92

1. 7	2. 9	3. 10	4. 12	5. 5
6. 12	7. 12	8. 12	9. 10	10. 9
11. 11	12. 3	13. 12	14. 7	15. 8
16. 7	17. 11	18. 8	19. 1	20. 0
21. 5	22. 11	23. 1	24. 2	25. 0
26. 12	27. 5	28. 1	29. 8	30. 11
31. 8	32. 9	33. 0	34. 2	35. 9
36. 7	37. 9	38. 1	39. 2	40. 10

Answer Key # 93

1. 2	2. 2	3. 3	4. 6	5. 3
6. 0	7. 12	8. 0	9. 0	10. 4
11. 6	12. 9	13. 12	14. 7	15. 5
16. 3	17. 5	18. 10	19. 12	20. 5
21. 5	22. 4	23. 7	24. 11	25. 8
26. 6	27. 5	28. 12	29. 7	30. 10
31. 3	32. 3	33. 0	34. 2	35. 5
36. 3	37. 9	38. 12	39. 11	40. 4

Answer Key # 94

1. 8	2. 3	3. 1	4. 4	5. 11
6. 5	7. 11	8. 6	9. 2	10. 6
11. 2	12. 0	13. 12	14. 5	15. 8
16. 9	17. 6	18. 6	19. 7	20. 1
21. 10	22. 11	23. 4	24. 2	25. 4
26. 8	27. 3	28. 2	29. 0	30. 7
31. 4	32. 5	33. 7	34. 10	35. 7
36. 3	37. 7	38. 10	39. 5	40. 0

Answer Key # 95

1. 11	2. 1	3. 12	4. 3	5. 12
6. 0	7. 8	8. 11	9. 11	10. 2
11. 9	12. 1	13. 11	14. 10	15. 4
16. 8	17. 1	18. 1	19. 8	20. 3
21. 1	22. 11	23. 4	24. 6	25. 7
26. 4	27. 8	28. 10	29. 4	30. 3
31. 1	32. 5	33. 6	34. 0	35. 1
36. 8	37. 10	38. 9	39. 8	40. 9

Answer Key # 96

1. 2	2. 11	3. 6	4. 10	5. 12
6. 1	7. 1	8. 10	9. 6	10. 9
11. 2	12. 5	13. 12	14. 9	15. 0
16. 9	17. 7	18. 12	19. 6	20. 4
21. 12	22. 10	23. 0	24. 7	25. 12
26. 0	27. 9	28. 0	29. 9	30. 3
31. 1	32. 11	33. 8	34. 3	35. 10
36. 8	37. 10	38. 4	39. 7	40. 12

Answer Key # 97

1. 1	2. 6	3. 4	4. 4	5. 5
6. 6	7. 11	8. 4	9. 1	10. 6
11. 7	12. 12	13. 2	14. 5	15. 8
16. 1	17. 7	18. 11	19. 9	20. 6
21. 4	22. 11	23. 4	24. 8	25. 7
26. 11	27. 9	28. 4	29. 11	30. 1
31. 10	32. 11	33. 12	34. 12	35. 5
36. 4	37. 2	38. 8	39. 4	40. 1

Answer Key # 98

1. 0	2. 3	3. 7	4. 5	5. 5
6. 1	7. 8	8. 5	9. 11	10. 9
11. 2	12. 3	13. 0	14. 6	15. 10
16. 8	17. 6	18. 1	19. 0	20. 2
21. 1	22. 0	23. 7	24. 12	25. 12
26. 2	27. 10	28. 3	29. 11	30. 1
31. 9	32. 3	33. 5	34. 9	35. 7
36. 5	37. 8	38. 10	39. 12	40. 7

Answer Key # 99

1. 1	2. 9	3. 8	4. 11	5. 10
6. 2	7. 2	8. 6	9. 2	10. 2
11. 2	12. 6	13. 3	14. 3	15. 1
16. 10	17. 6	18. 0	19. 10	20. 2
21. 7	22. 5	23. 7	24. 0	25. 10
26. 12	27. 12	28. 9	29. 3	30. 11
31. 4	32. 12	33. 9	34. 5	35. 7
36. 9	37. 4	38. 0	39. 4	40. 6

Answer Key # 100

1. 0	2. 9	3. 11	4. 8	5. 6
6. 10	7. 0	8. 5	9. 0	10. 5
11. 3	12. 8	13. 3	14. 8	15. 10
16. 2	17. 3	18. 6	19. 7	20. 12
21. 2	22. 7	23. 8	24. 5	25. 4
26. 8	27. 3	28. 3	29. 12	30. 0
31. 9	32. 9	33. 3	34. 10	35. 7
36. 2	37. 12	38. 1	39. 9	40. 3

Track your progress

Set	Time	Score
1		
2		
3		
4		
5		
6		
7		
8		
9		
10		
11		
12		
13		
14		
15		
16		
17		
18		
19		
20		
21		
22		
23		
24		
25		

Set	Time	Score
26		
27		
28		
29		
30		
31		
32		
33		
34		
35		
36		
37		
38		
39		
40		
41		
42		
43		
44		
45		
46		
47		
48		
49		
50		

Set	Time	Score
51		
52		
53		
54		
55		
56		
57		
58		
59		
60		
61		
62		
63		
64		
65		
66		
67		
68		
69		
70		
71		
72		
73		
74		
75		

Set	Time	Score
76		
77		
78		
79		
80		
81		
82		
83		
84		
85		
86		
87		
88		
89		
90		
91		
92		
93		
94		
95		
96		
97		
98		
99		
100		

More math workbooks available from EDventure Learning

Made in United States
North Haven, CT
05 November 2026

21777130R00061